Florence Caddy

Through the Fields With Linnæus

A chapter in Swedish history. Vol. I

Florence Caddy

Through the Fields With Linnæus
A chapter in Swedish history. Vol. I

ISBN/EAN: 9783744743419

Printed in Europe, USA, Canada, Australia, Japan

Cover: Foto ©ninafisch / pixelio.de

More available books at **www.hansebooks.com**

A CHAPTER
IN SWEDISH HISTORY

VOL. I.

PRINTED BY
SPOTTISWOODE AND CO., NEW-STREET SQUARE
LONDON

STATUE OF LINNÆUS AT STOCKHOLM

THROUGH THE FIELDS WITH LINNÆUS

A CHAPTER IN SWEDISH HISTORY

BY

MRS FLORENCE CADDY

AUTHOR OF 'FOOTSTEPS OF JEANNE D'ARC' ETC.

IN TWO VOLUMES—VOL. I.

ON THE CASTLE HILL, UPSALA

LONDON
LONGMANS, GREEN, AND CO.
1887

TO MY HIGHLY ESTEEMED AND VALUED FRIEND

DR HENRY TRIMEN, F.L.S. &c.

Director of the Colonial Botanical Establishment, Ceylon

WHOSE KIND ASSISTANCE HAS ENABLED ME WITH THE

MORE CONFIDENCE TO OFFER TO THE PUBLIC

THIS SKETCH OF THE LIFE AND TRAVELS

OF THE GREAT NATURALIST

CONTENTS

OF

THE FIRST VOLUME.

CHAPTER	PAGE
I. THE LINDEN TREE OF LINNHULT	1
II. WEXIO	27
III. LUND UNIVERSITY	54
IV. UPSALA	85
V. DEAN CELSIUS COMES	114
VI. THE NORTH SWEDISH PROVINCES	136
VII. 'LACHESIS LAPPONICA': JOURNEY ROUND LAPLAND, 1731, MAY TO NOVEMBER	168
VIII. ROSEN VICTOR	198
IX. ITER DALECARLIUM—THE FAIR FLOWER OF FALUN	213
X. TAKES HIS DOCTOR'S DEGREE IN HOLLAND	259
XI. LEYDEN—THE FAT OF THE LAND	291
XII. A VISIT TO ENGLAND	318

ILLUSTRATIONS IN VOL. I.

STATUE OF LINNÆUS AT STOCKHOLM . . *Frontispiece*

ON THE CASTLE HILL, UPSALA . . *Vignette title*

LINNÆUS IN SMÅLAND . . . . . *To face page* 22

MAP.

(*At end of volume*)

THE TRAVELS OF LINNÆUS, 1735-38

THROUGH THE FIELDS WITH LINNÆUS

CHAPTER I.

THE LINDEN TREE OF LINNHULT.

> Beneath yon birch with silver bark
> And boughs so pendulous and fair,
> The brook falls scattered down the rock,
> And all is mossy there.—COLERIDGE.

AT Råshult, in the heart of Småland, a province of South Sweden, on a slope beside the trunk railway line, stands a small shingle-roofed wooden house, painted deep red, with white windows draped with the whitest muslin the best laundry ever aided a bleaching-ground to produce. A granite obelisk before the house, between it and the rail, tells all the world, or, to be accurate, the few persons who daily travel through Småland by the slow cattle and timber train, that Carl von Linné, oftener spoken of as Linnæus, was born here on May 23, 1707. The obelisk was erected in 1866. No other building is visible until we arrive at the small station at Liatorp.

This is, indeed, being flung into *medias res*, as Horace recommends; for if any place can fairly be defined as *medias res*, it is Råshult in Småland. Even in Sweden there is no getting at it without patience. It is the kernel of everything; leading apparently from nowhere to nowhere, yet really on the main road to every town of interest in Sweden.

At 10 P.M. the late and lingeringly slow 'cargo' train stops at Liatorp station for 'Linnæus,' as every one understands and is careful to inform us, and moonlight on the islanded lake Möckeln, and the last gleams of dying daylight, at the end of May, make it easy to find the small hotel with its '*Rum för Resande*.'

We can see the mat of fresh spruce boughs laid, as is customary, at the foot of its wooden threshold steps, and above this we meet the welcoming smile of a pleasant-looking woman, who has been sitting in the porch watching by the tender light, still tinting the sky with daffodil and wild-rose colour, to see the train come in.

'Yes, this is right for Linnæus. And so the stranger ladies want to find out all about our Linnæus? This is charming. Yes, here is Stenbrohult, his early home; and yonder is Råshult, his birthplace. To-morrow you can drive over and visit it. It is about half a mile from here.'

'It is not worth while to drive that little step.'

'Half a Swedish mile!'

[1] *Rum* is a room in Swedish; it is pronounced room, and not rum.

'Ah, to be sure—there is a difference.'

'So the English people really care about our Linnæus? This is delightful.'

It seems very amusing to them likewise. The train was still waiting, and would have carried us on farther had we been misled.

Meanwhile mistress and maid draw out a bed which shuts up telescope-fashion at the foot, and prepare a sofa whose cover lifts off, and, drawing out at the side, forms a trough-like receptacle for an extra guest. All sofas, however splendid the apartment, are thus formed as spare beds in Sweden. Going to bed here is like laying oneself by in a drawer. Presently the round table, sleek with white linen, is spread with a star-shaped arrangement of tiny glass dishes of relishes, served to provoke appetite (they do this with hungry English people), to be removed by a second course of exquisitely cooked cutlets and potatoes—a pleasant sight at 11 P.M. for famished travellers weary with the time-murdering train, that dawdles fifteen minutes at every wayside station, where there seems no excuse for having a station at all; and, besides these, a junction at every alternate station gives one ample time for making excursions in the district. The groaning train of timber-trucks stretching from horizon to horizon, with some pig and cattle vans, and a passenger carriage at the end of all as a concession, steams slowly in and out of each station at one-horse—no, one-donkey or puppy-dog power. Let not over-wise travellers avoid the snäll tåg[1] (snail train) lest a worse thing, the cargo train, befall them. But here we are on our ground, in *medias res*, as explained before, which is a comfort; for I defy anyone to get at Stenbrohult from any civilised place, other than Scandinavian, within a fortnight or ten days at least.

Liatorp is in the parish of Stenbrohult, of which Nils Linnæus, father of our hero, was rector, so this is the scenery of the great botanist's early life.

These wooden houses, the only sort known in Swedish villages, are much larger than from the outside they appear to be. They have shingle roofs, set on mostly at a right angle—sometimes more, sometimes less, but thereabout. This seeming trifle shows a naturally inartistic feeling in the Swedes. The right-angled gable always makes a common-looking house; it has no specific character.

The wooden walls being so thin makes the apartments surprisingly roomy inside; yet they are warm and comfortable withal. Every bedroom by day becomes a fine sitting-room with handsome mahogany sofas, even in poor houses. The flowers, too, are a great adornment, for the inhabitants keep plants in every window. That cactus (you would take it for a green rock or a fossil cactus) is seventy years old, for certain; it has been so long in the family, and was not young when it came to them. It may be over a hundred years old. It or its immediate ancestor belonged to Nils Linnæus. It is not rash to suppose it one of

[1] *Schnell-zug*, fast train.

his four hundred rare plants; though more probably it was a Mexican importation, sent by Linnæus to his father. I do not say it was so, but much of the best legal evidence is made up of likely conjecture. The other plants here are flowers—roses and such-like—with less pedigree but more beauty—this is *la vieille roche.*

The smallest cottages here are comfortable, and the people, though poor in actual coin, are yet easy, happy, and contented. One can best judge of the happiness of a country by the condition of its poor people. Here, though it is hard to make a living, there appear to be no poor people in our sense of the word; that is, none verging on pauperism. The villages are trim and clean, without being over scrubbing-brushed as in Holland. The floors are of clean bare boards.

Give your linen to the maid, and you will see her wash it at the pump, soap it, beat it on a bench thoroughly with a kind of cricket-bat, bleach it on the flowery turf, and return it to you lint-white, with all the patent washing-powders and dirty messes, with which townspeople give themselves blood-poisoning and all manner of skin diseases, driven out of it. Brave, strong girl; I see her in the garden doing her washing so honestly. The farmyard with the fowls, and, besides the pump, a well with a bucket to wind up, has a neatness, without being at all prim or fancy-farmed, which makes it very pleasant to look down upon. A long ladder reaches to the attics or garrets, where the swallows build in the eaves. There are also wooden steps up to the slightly elevated ground floor.

Many vehicles, of a sort between a cart and a carriage, drive into the yard, and the drivers unharness and put up their horses in free, old-fashioned, homely style, and, doffing their huge frieze overcoats, awaken themselves up into the belief that it is a pleasant summer-like day. Such a funny old equipage has just driven into the yard! an ancient form of cabriolet or chariot with a hood; low, small, crunched-up, and, oh, so shabby-genteel! Out of it step an ancient pair, in clothing like old pictures, just spoiled with a few modern remnants of the fashion of ten years ago. The lady trails her snuff-brown silk skirt, with one scanty flounce at the bottom of it, through the farmyard with a genteel amble, keeping her quality from contact with the general coffee-room. She is unaware that dresses are worn short now, and that flounces such as hers are no longer admired. The gentleman's tailor can never have smiled again after executing that esteemed order. The gentleman, furnished by him, is short, and stout, and brown of extreme neutrality—a faded brown, further neutralised by long lying by, when the style of the day—I speak of Sweden—is a symphony in spinach colour, with velvet collar of a livelier hue of green, a bronze-green billycock hat with a peacock's feather in the band, and a tasty alpaca umbrella of ultramarine blue, all cushiony and full. They use a simple sort of sledge here in the snow-time—one sees them lying by in handy corners; in summer they run light home-made well-made carriages, constructed to hold two medium-sized and two small persons, with cushions of home-grown and home-tanned hides. One of these was now brought round to take us to Råshult.

The sandy road to Råshult meanders about the railway-line like the serpent round the rod of Æsculapius. The railway men ply swiftly to and fro between the stations on the line, paddling their little tricycle trucks, or six of them more arduously pumping along their 'Sociables.' This is much quicker travelling than the train.

Whortleberries, juniper shrubs, and wild strawberries form the undergrowth beneath the pines and among the grey boulders set in wood anemones, among which as we passed lay a snake curled up like one of the twisted cakes used as the sign of a baker's shop. Three flaxen-haired, dark-blue-eyed girl children dressed in shades of pink and grey and rosy scarlet greeted us from their cottage garden gate with wondering but modest gaze.

The people are polite, the wayside greetings are very courteous, yet everyone minds his own business, and a pushing crowd never gathers round an artist as in Belgium and elsewhere. There is no fear of pickpockets, or other robberies or disagreeables.

I will here give part of Linnæus's own characteristic description of the scenery, taken from his diary.

'Stenbrohult, a parish of Småland, is situated on the confines of Skåne, in a very pleasant spot adjoining the great lake Möklen, which forms itself into a bay about a quarter of a (Swedish) mile long, and in the centre of this bay stands Stenbrohult church. It is surrounded on all sides, except to the west, where it fronts the lake, by well-cultivated lands. At a little distance to the south the eye is relieved by a beech wood; to the north the lofty mount Taxås[1] rears its head, and Möklen lies on the opposite bank of the lake. Moreover, to the east the fields are encompassed with woods which westward inclose broad meadows and large spreading trees. In short, Flora seems to have lavished all her beauties on the spot that was to give birth to our botanist.'

We drive through an avenue of hoary-lichened firs with the lake Möckeln shimmering between their stems, coloured no longer with the glassy reflections of last night before the sundown, but fresh, blue, and sparkling in the limpid air, fragrant with flowers and buds of the lichen-clothed juniper shrubs. We cross a bridge over an arm of the lake. The influence of a gardener lasts long and spreads wide; we still perceive the influence of Nils Linnæus and his clan, all of them great gardeners, in the variety and comeliness of the vegetation, which is hereabouts unusually rich. Yet one tree that used to flourish here, the famous lime tree of the Linnæi, is conspicuously absent. There is no lime tree growing here now, or none of any stateliness. And yet there might be; for if De Candolle's list of the ascertained

[1] Linnæus had seen few mountains when he wrote this. It is as if we might say, 'the lofty Primrose Hill rears its head,' &c.

ages of certain trees be correct, there are limes that have lived a thousand and nearly twelve hundred years.[1]

This famous lime tree, according to Pulteney, one of the most careful biographers of Linnæus, stood on the farm where he was born, and three progenitors of his family took their names from it—Lindelius, Tiliander, and Linnæus. This shows the inaccuracy of even a careful historian, for Linnæus was not born on a farm at all, but at the parsonage house of Råshult, and his ancestors who named themselves from the lime tree lived at Hwitaryd, near by. It is not unusual for Swedish families to name themselves from natural objects.

The peasants regard the lime tree as sacred; in early spring they deck the graves of lost relatives with its fresh green boughs.[2]

A large linden tree would always be an object of note in that land where the pine, the spruce fir, and the birch are the principal vegetation above the variegated carpet of the ground. The tree in question may be here, should be here, but I have not identified it, nor could I hear of it. The tradition of its three branches dying at the extinction of the three families and the dead stump remaining is, I suspect, a legend.

For those who care for the study of race, of descent of talents and qualities through pure genealogies, the

[1] The lime is one of the most lasting of trees, living to 1,076–1,147 years. This is measured by the concentric zones. Professor Henslow considers De Candolle overrated the ages of his trees one-third.

[2] Horace Marryat.

story of the linden tree of the Linnæi should have great interest, showing as it does how the pride of hard-won place went hand in hand with a deep and increasing love of nature, inherited from three worthy peasants of Hwitaryd (who dwelt under the shadow of the great lime tree) whose descendants intermarried, and their fine qualities combined to form one brilliant descendant, the flower of the family tree—the splendid Linnæus.

For all that Buckle holds that there is no such thing as hereditary transmission of qualities, no virtue in pure race, the general experience of the world runs otherwise.[1]

In speaking of a botanist like Linnæus it is incumbent on one to mention root and branch, and Linnæus was proud of his genealogy. In his notes made for his autobiography—which never became a book—he gives in full the genealogy of the Linnæi, with their botanical and clerical traditions, which I shall epitomise here. Skip it, ye who care not for such matters; easier reading lies beyond. Yet one has to learn less interesting lines of kings, and there are crowds who read the pedigrees of horses in the stud-book and racing calendar.

'Ingemar Suensson, a peasant at Jomsboda, in the parish of Hwitaryd in Småland: from him descended Charles Tiliander, who took his name from a tall tilia standing between Jomsboda and Linnhult. He studied

---

[1] Galton, in his book on Hereditary Talent, says, 'I would strongly urge that the sketch should be pretty exhaustive as regards the nearer kinsfolk, male and female, certainly including aunts and uncles on both sides, and preferably great-aunts, uncles, and cousins. This has great statistical value.'

at Upsala in 1660, and died without issue 1697. His brother, Suen Tiliander,[1] studied at Upsala 1678. He lived as domestic chaplain to Count H. Horn at Bremen, and died rector of Pjetteryd in 1712.' [Our little Linnæus may have remembered him, his great-uncle. Suen's sister Ingrid married Carl's grandfather.] 'He was a peculiar lover of gardening and natural history. His sons were Abel Tiliander, who succeeded him as pastor, and was drowned in a well in 1724, and Nicholas, chaplain to a regiment. The latter left issue, Carl Tiliander, born 1701, who studied at Lund 1720, became adjunct teacher of Philosophy there in 1729, and adjunct teacher of Divinity there 1730.'

Doubtless this Carl, who was six years older than our Carl Linnæus, was held up as a model to his younger cousin, who was reckoned among the dunces. He was high in Lund University at the time Linnæus was entered there as a student, with a bad character from his grammar school. We do not hear of the two having had much communication. I fear me Carl Tiliander was a prig, and ashamed of his country cousin. Yet the Tilianders seem to have been the pedagogues of the Linnæus family for a long while, for Suen, the pastor of Pjetteryd, took our hero's father, Nils Linnæus, into his house 'to educate with his children, and, having a good garden, he gave him also a taste for horticulture'; and a certain John Tiliander, a severe man, which is all we can find out about him, was the earliest tutor of the

[1] Linnæus's maternal uncle.--SIR J. E. SMITH.

great botanist. Linnæus now takes the line of another stem of his family.

'Anders, a peasant at Jomsboda.' These peasant ancestors were all men of virtue and value. 'His progeny were: (1) Ambern Lindelius, born 1600; took his name from the same linden tree.' He was the first to do so. 'He was made Master of Arts in 1632, teacher, lecturer, and rector, &c., and died in 1684. (2) Lars Lindelius, who died rector of Jönköping in 1672. Eric Ambern Lindelius, son of the former, studied at Upsala 1655, died preacher at Quänberga in 1715,' when Carl Linnæus was eight years old. 'Lars Lindelius' son John was a physician of great repute at Wexio. He studied at Lund 1672, Upsala 1680, and died in 1711.' Thus both the Swedish universities and many of the rectories in South Sweden teem with Linnæan traditions. 'No remaining males of this family,' says Linnæus.

Now comes the line of Linnæus's father's family, the main stock.

'Benge Ingemarson, also a peasant in the parish of Hwitaryd, had issue Ingemar Bengtson, born 1633. He was farmer of the manor of Erickstad. His son Nicholas (Nils), who also took his name of Linnæus from the same linden tree, was born 1674; assumed clerical functions 1704.' At the age of thirty-one he married Christina Broderson, the young daughter of the pastor of Stenbrohult, who was only seventeen. This was in 1705; at the same time he was appointed vicar of Stenbrohult, in the curacy of the district of

Råshult. At the time of the great botanist's birth Nils Linnæus was comminister, which, on the Swedish Church establishment, is a clergyman somewhat similarly circumstanced to one who in England serves a chapel of ease.[1] The 'Swedish Biographical Dictionary' mentions many other ancestral connections and collateral branches, including a cousinship with the British Admiral Kempenfelt, who was drowned in the 'Royal George,' 1782, in the course of five pages of genealogical tables, comprising a number of respectable people, nearly all of them clergymen or medical doctors.[2] There is also a genealogical table in the appendix to Pulteney's biography, going further and more minutely into the pedigree, including various interesting particulars, such as this concerning Ingemar of Waras, in the parish of Hängeryd, who was blind many years, and spontaneously recovered his sight in advanced age. But this is enough for the indulgent reader.

GENEALOGY OF THE LINNÆUS FAMILY.

LINDELIUS.	TILIANDER.
Anders, a peasant at Jomsboda in the parish of Hwitaryd, Småland, had sons—	*Ingemar Suensson*, a peasant at Jomsboda, had sons—
1. *Ambern Lindelius*, born 1600, died 1684.	1. *Carl Tiliander*, died 1697.
2. *Lars Lindelius*, died 1672.	2. *Suen Tiliander*, died 1712; and a daughter, *Ingrid Ingemarsdotter*.
Ambern's son—	
Eric Ambern Lindelius, died 1715.	Suen Tiliander had sons—
	Abel Tiliander, pastor, drowned 1724.
Lars' son—	*Nicholas Tiliander*, army chaplain.
John Lindelius, physician at Wexio, died 1711.	Nicholas's son, *Carl Tiliander*, born 1701.

---

[1] Pulteney.   [2] *Svenskt Biografiskt Lexikon.*

LINNÆUS.

Benge Ingemarson, peasant in the parish of Hwitaryd, had a son,

{ *Ingemar Bengtson*, born 1633, farmer of the manor of Erickstad; he married
{ *Ingrid Ingemarsdotter*, sister of Suen Tiliander, pastor of Pietteryd.

They had a son,

{ *Nicholas (Nils) Linnæus*, born 1674, rector of Stenbrohult; he married
{ *Christina Brodersonia*, daughter of his predecessor in office.

They had two sons and three daughters—

Carl Linnæus, born 1707, married Sarah Elizabeth Moræa; had two sons (both died childless) and four daughters.

Anna Maria, born 1710, married Gabriel Hök, rector of Wirestad.

Sophia Juliana, born 1714, married Johan Collins, rector of Rÿsbÿ.

Samuel, born 1718, married the daughter of the prebendary of Markaryd; had several daughters, no sons.

Emerentia, married — Branting, receiver of the land-tax in the Hundred of Sunnerbo.

No heir male of the three families. The arms of Von Linné were broken on the tomb of Carl von Linné, son of the great Carl Linnæus, ennobled as Von Linné.[1]

Nils Linnæus afterwards became rector of Stenbrohult. His father-in-law, Samuel Petri Broderson, rector of Stenbrohult, died December 30, 1707, of a fall by which his clavicle was broken. The vicar of Wexio succeeded Samuel Broderson, but, dying in the same year, Nils Linnæus succeeded him in the living of Stenbrohult.

The family tree is the linden of Linnhult.

Thus Linnæus was, in fact, rooted in Råshult; it was more than an ordinary birthplace. The linden tree

---

[1] How often we see in cases of great hereditary ability the line dies out after the most talented member of it has brought it into special prominence.

under whose shade the family grew up stood in the vicinity of his native place, between Jomsboda and Linnhult. The linden tree has passed away, but in this cottage, this very cottage, Linnæus's father dwelt, so long do these wooden houses last. This one looks as new and strong as do the other houses round, and as cheerful with its white muslin window draperies, for it is inhabited, and the climbing plants growing up round it, ever youthful in their buds and blossoms. Children still peep from those windows, still play about this sloping garden. A little pair are before me now. But that this boy's eyes are of the usual Swedish blue, like the speedwell of their fields, in this fair child I can almost imagine I see the intelligent and bright-faced Carl Linnæus, a boy with rosy cheeks, sparkling brown eyes, and light silky hair, almost white in its fairness; and that tiny maiden, with the dazzlingly fair neck, and flaxen locks escaping from under her cotton gipsy bonnet, might be the little Anna Maria Linnæa, long since lying in respected sleep as Fru Hök in the rectory churchyard of Wirestad, near by. Nils, the perpetual curate of Råshult, having been born in 1674 makes the little house connect us with that date, which has so long since drifted into history, in a more intimate way than do many more ancient buildings. Life here altogether carries us back in the past, so completely is it the life of Linnæus's own day and that of his ancestors before him.

There has been no regular biography written of

Linnæus since Stoever (of Altona) wrote in German his valuable 'Life of Linné' in 1794, and **Pulteney his in English in 1805.**[1]

These two biographers abuse each other politely in long prefaces. Stoever says of Pulteney's book, 'It is in several other respects imperfect and deficient. The learned author ought to have had recourse to Baron Haller's "Bibliotheca Botanica," tom ii. What follows is a translation of *this* work.' Pulteney speaks of Stoever's 'Life of Linné' as containing interesting particulars, 'but it is not without a considerable number of errors.'

Sir William Jardine, in his brilliant epitome of both books, made as a short biographical notice of Linnæus for the Naturalists' Library, speaks of Linnæus's diary [2] as owing its preservation to Dr. Maton who edited it.[3] Almost as precious as this are the letters and diaries of travel kept by Linnæus, which came with his other collections into the possession of Sir James E. Smith, the founder of the Linnæan Society. These papers, written either in Latin or Swedish, have been partially edited and translated by him, and some few of the diaries have been separately published in German, but some of them have never hitherto been brought to

---

[1] Turton's biography, written in 1806 to accompany a translation of Linnæus's General 'System of Nature,' is compiled from these. The smaller biographies are abridgments of Stoever.

[2] The marvel is that Stoever did his work so well without the diary and documents that Dr. Maton appended to the second English edition of Stoever.

[3] The diary, down to 1730, was put into Latin by Archbishop Menander. It is written in the third person.

light at all. Sir J. Smith is our most trustworthy authority on this subject, as he possessed materials of which both Stoever and Pulteney were ignorant, although he only used them biographically in a short memoir written for Rees's 'Cyclopædia.'

What the present generation knows of Linnæus is an obsolete system and a few trivial anecdotes. In painting his portrait I have tried to give as a background the things he saw, the scenes he moved in, the continuous diorama of his life, which abounded with adventure more than usually falls to the lot of scholars 'whose fame is acquired in solitude.' I wish it may be thought a pleasant yarn about Linnæus.

Stoever, and all the short biography writers who about his time pillaged rather than translated him, begin with a hot dispute concerning Linnæus's birthday. Some say it was the 3rd of May, some the 13th, some the 23rd, and various other dates.[1] Linnæus in his genealogical table says: 'On May 12, 1707, at Råshult in Småland, was born Carl Linnæus'; but as his own flowery language in his commenced autobiography says he was 'brought into the world in a delightful season of the year, between the months of frondescence and flo-

---

[1] The New Style being then in process of gradual adoption in Sweden, the year 1704 was regarded as a common year in that country; consequently the true date of Linnæus's birth, according to our present reckoning, was May 23, 1707; the commonly received date, May 24, being an error due to supposing the calendar in Sweden and Russia at that time to be identical.—*Encycl. Brit.*, JACKSON.

rescence,' this gave a good opening for controversy. The 'Times' of April 1885, speaking of the new statue of Linnæus in the Humle Garden at Stockholm, says: 'On the 13th of next month a statue of the celebrated Swedish botanist Linnæus will be publicly unveiled at Stockholm. The day will be the 178th anniversary of his birth.' The obelisk says he was born on May 23. Linnæus's own diary fixes the date with scrupulous exactness, as May $\frac{12}{22}-\frac{13}{23}$, between 12 and 1 in the night.

The reason of these aberrations regarding his birthday is that, taking the 13th of May as about a central date, some authors in their 'cuteness, thinking they are the first to remember the fact of the late change in Sweden from the Old to the New Style in the calendar, have put him on or back eleven days, or some only ten days inclusive, not being able for the life of them to remember whether it should be eleven days forward or backward; accordingly the date ranges from the 3rd to the 24th of May, an important difference in the short Swedish spring. Carl was the first-born child of his parents; other little ones followed quickly on—three daughters and a second son.

We will have one short look round the curate's cottage at Råshult before removing with him to the somewhat larger house and much larger garden of Stenbrohult rectory. The granite obelisk, surmounted with the Polar star in gold, stands tall in front of the small red cottage (of the curacy) at the top of the sloping garden, still set with flowers and beehives on grass thick with anemones and tiny pansies, and wild strawberries under the yellow-flowering gooseberry bushes and other shrubs. The cottage stands in a beech grove which forms with the spruce, larch, and other trees a forest around it. The ground at the foot of a good-sized oak growing below the cottage is powdered with wood anemones. The land is undulating hereabout, and very agreeably broken and diversified. The railway-line passes directly before the obelisk and cottage, being only divided from the garden by a pair of iron gates. Beyond a stream, which one passes by a plank bridge, a wood rises on the opposite side across the railway.

The well is still worked by a pole lever, one of the earliest and simplest ways of raising water for gardening purposes. Above the vibrating sound of the woodpecker's tapping rises the prolonged coo of the woodpigeon. The air is vocal with birds and perfumed with buds and flowers. It is the very fittest early home for a student of natural history, the science of peace. The garden is walled on two sides with granite, the large stones being smoothly laid and fitted without mortar. A granite slab forming a small table in an arbour is inscribed with Linnæus's name; the Polar star and other devices are decipherable on it, traced in outline with tinges of colour. This, of course, has nothing to do with the infancy of Linnæus.

His father was appointed rector, instead of curate, of Stenbrohult in 1708, and the family moved to the rectory when Carl was a year old. The present rectory house is not the home of Linnæus's childhood—that was burnt down some forty years later and rebuilt. Here was a much larger garden, and Carl was as a child in the Garden of Paradise. 'From the very time that he first left his cradle,' says the enthusiastic Turton, 'he almost lived in his father's garden, which was planted with the rarer shrubs and flowers; and thus were kindled, before he was well out of his mother's arms, those sparks which shone so vividly all his lifetime, and latterly burst into such a flame.' 'The same thing that is said of a poet—*nascitur, non fit*—may be said without impropriety of our botanist.'[1] Carl was nursed in beauty, fragrance, and pure delights. His toys were flowers, and Christina, his young mother, herself with only eighteen years of youth, used to stop his cries by giving him a flower to play with.

The smallness of the rector's income obliged him to make the best of husbandry. He was his own gardener. His child was his constant companion, enjoying to the full

> Delight and liberty, the simple creed
> Of childhood, whether busy or at rest.[2]

Here on the scene one seems to see the sunny-haired child running about among these ferny foregrounds, his baby feet—sometimes bare like his young brethren around, sometimes, as became the rector's son, with tiny canvas shoes with a buckle-strap across the instep—care-

---

[1] Linnæus's Diary.   [2] Wordsworth.

fully stepping between the plants so as to injure none of them, as I have seen a little London-bred boy do among the ferns at Hampstead Heath. Perhaps Carl oftenest wore those national thick shoes with heels, but cut off low at heel, like slippers, which it is almost impossible to wear out. In these he went out scampering and chasing the butterflies across the broken ground where the granite boulders are almost lost among the whortle-berries, his keen eyes and swift feet following their flight into the grey mystery of the fir woods until tempted away by the discovery of a nest of game birds, all full of dear little palpitating balls of fluff.

His eyesight was from a child remarkably acute—a seemingly indispensable requisite to the naturalist, did we not remember that Huber, Reboul, and Rumphius, among the most eminent observers of nature, have been blind.

Carl needed his keen bright eyes to trace the airy path of the butterflies, for, as a rule, these are very small in Sweden.[1] The brimstone butterfly, the only large one I saw here, is one of the largest, at least among the common ones.

I have just caught, killed, and stuck a tiny fairy, a blue butterfly,[2] smaller and more fragile than our smallest chalk-hill blues. Did I call natural history the science of peace? Oh, monstrous fiction! But Linnæus was as yet innocent of trying to compass their death. He wanted to keep them alive, as most

---

[1] The family of the *Lycænidæ* being numerously represented.
[2] One of the *Lycænidæ*.

children do; it was only by sad accident that they were bruised and maimed. He formed a museum of live insects. He scarcely knew which he loved best, the caterpillar or the plant it fed upon. He loved the troops of ants that crowd the dust with palpable life. Some of the sorts found hereabout are large enough to be visible to the most casual observer.[1]

What pleasure to a child thus to run, beloved yet wild and free, beneath the trees, sheltering the cooing doves, and the dryad's hair of the silver birch, his feet lapped in leaves of the wild lily of the valley with miniature racemes [2]; the pimpernel and fern curls fringing the foundations of the boulders, which served little Carl for seats and tables! Where we can generalise only a mazy bewilderment of grey stems, and in the foreground a crumbling grey intricacy of boulders touched with orange lichen, the colours of the orange-tipped butterfly, his classifying infant eyes can spy the minute green butterfly,[3] invisible to anyone else, upon the whortleberries the moment it waves the brown upper sides to its wings preparatory to a fresh start. It has been safe, even from Carl, so long as the metallic green under-sides to its closed wings hid it among the crowd of leaves. Distinct to him also are a small brown-and-white speckled butterfly, and an atomy dark brown spotted one,[4] nearly black; and among the commoner

---

[1] There are five species of *Formica* found in Sweden. One kind is said to be eatable.

[2] *Maianthemum bifolium.*   [3] *Thecla Rubi*, Lin.

[4] *Cænonympha Hero*, Lin.

LINNÆUS IN SMÅLAND

white cabbage butterflies he can discern a rarer fairy, a tender Psyche, looking to the ordinary world like a wood anemone or an oxalis flower.[1] The brimstone butterflies are a thought yellower than ours, and there are many small-sized tortoiseshells. George Eliot mentions how not one polype for a long while could Mr. Lewes detect in some seaside holiday ramble, after all his reading—so necessary is it for the eye to be educated by objects as well as ideas. For one thing, however, the little Linnæus's eyes were never strained over the horrible uncivilised print that the Germans blind themselves with, as, though the Danes and Norwegians use the Gothic character, the Swedes use Roman letters, and the Swedes seldom wear spectacles.

The church of Stenbrohult is three-quarters of a Swedish mile from Liatorp station, and the parish is very scattered, entailing considerable labour on the clergyman; but the congregation make light of twelve English miles to go and come to church. The village and the indispensable all-sorts shop of *Diverse Hökeri*, and the *Bageri*, or baker's shop, are at Liatorp. There are no squalid cottages in Liatorp such as we too often see in Devonshire villages; yet all nature is less kind, and the winter is cruel. Though in the stony wilderness of Småland there would be hardly a square yard of turf unless cleared by hand labour, and they cannot plant a cabbage till they have cleared a space, the cheerful content of the people would be surprising, but for the thought that a field once cleared

[1] *Leucophasia sinapis* of the *Pieridæ*.

is always clear and profitable when once the boulders are piled up into fences. If I seem tediously minute, remember that Stenbrohult is the foundation stone of Linnæus's history. This is the summer life. In winter the social joys are perhaps keener and more common, for it is easiest to get about when frost has made the land and water all alike, when the snow has filled up and smoothed the roughnesses of the ground between the boulders, and the Swedish landscape is daylighted by its own purity of snow, its landmarks effaced, and one can best travel by the compass of the stars. The worst season is before the frost sets in, when these Northern forests have a dreadfully aguish feeling. The little Linnæus, all his life a gouty subject, was very susceptible to neuralgia and suffered much from toothache.

His pleasures, however, as a child redeemed his pains. He has recorded how the love of natural science that followed him through life was first decidedly displayed when he was scarcely four years old. The child was indeed father to the man. It must have been at Whitsuntide after his fourth birthday when Carl accompanied his father to a feast at Möckeln, on the other side of the tall alder-fringed lake. In the evening the guests seated themselves on some flowery turf, listening to the pastor, who explained to them the names and properties of various plants, showing them the roots of the Succisa, Tormentilla, Orchides, &c. The child paid deep attention to all he saw and heard, and from that time never ceased harassing his father with questions about the name, qualities, and nature of every plant he met with; but though his memory was good, and later became remarkably so from its constant exercise with attention, childlike he forgot the names of the plants and the result of all his questions. These things require to be impressed on children's memory by constant repetition, line upon line, like the nursery rhymes and other lore they learn. His father refused to tell him more until he showed with the curiosity a determination to remember. Indeed, it was a long lesson to learn the names of all the plants in the home garden; for Linnæus in a letter to Baron Haller says it contained more than four hundred species, many of them rare and exotic.[1] His father was his tutor in other things besides natural history. He taught Carl Latin, religion, and geography, 'to qualify him for the pulpit and to conduct his botanical studies more skilfully,' until he was seven years old, when he was placed under the care of John Tiliander, a relative, who, I suppose, came to stay with the family, or came, perhaps, as curate.[2] He was in no way fitted to be tutor of an intelligent, vivacious, and peculiar child, the child of the young century, John having in teaching but one idea of his own—an idea already antiquated: that of obstinate severity—an idea impossible to maintain at a time when

---

[1] 'To this early discipline Linnæus afterwards ascribed his tenacious memory, which, added to his sharpness of sight, laid the foundations of his eminence as a reforming naturalist.'—JACKSON.

[2] Linnæus marks the date Sept. 15, 1714.

the swathes and bandages of the European intellect were bursting. The budding energies might be controlled and swayed, but could not at that impulsive epoch be sternly repressed. Childhood has the mind in miniature, but, having its seed leaves still on, we do not always recognise its sort.

To the late day of his writing his autobiography Linnæus bore a grudge against John Tiliander. This man, morose, and probably disappointed in the ambition he shared in common with his family, vented his discontent with circumstances on poor little Carl.

But the boy's pains and penalties were mitigated by a joy, a new sense—that of proprietorship. At eight years old his father allotted Carl a piece of ground of his own, and he at once began to form a botanical garden in miniature on an independent plan. His love of science disturbed even his father when the boy brought in weeds and wild herbs hard to eradicate, and worse than these—wild bees and wasps, with all their concomitant inconveniences. In a garden necessarily cultivated for profit such practical science had to be discouraged.

CHAPTER II.

WEXIO.

> The breeze that flung the lilies to and fro, and said
> The dawn, the dawn, and died away,
> Thence shall we hear
> The music of the ever-flowing streams,
> The low deep thunders of the booming sea.
> *Clouds*, ARISTOPHANES (translated by MR. COLLINS).

IN the spring of 1717,[1] Carl's father took him to Wexio to be entered at the grammar school (or trivial school, as Linnæus calls it in his diary) in that town where already his connection John Lindelius had been celebrated as a physician. It is true John had been dead these six years, but his interest would still be alive in the place, and might be useful to the boy now for the first time leaving his parents' roof. Carl's outward appearance had been transformed to suit his altered circumstances. The long silky white hair was cut short as befits a schoolboy, and he was provided with new boots for state occasions—high loose boots reaching up the calf. Linnæus's account of his schooldays is long-worded, after the fashion of his time; but Stoever's biography of him is very funny. He begins the story

---

[1] A note-book of Linnæus's says September 1714, but this date is evidently an accidental interpolation. See note 2 p. 25.

of Carl's going to school, 'At the epoch of this determination'—viz. of his father sending him to school—'Linnæus had seen his second lustre.' This makes Stoever difficult reading, one so often stops to laugh or to consult the dictionary (often Lemprière). His full-sailed prose expresses everything in long measure. Stoever's rococo pages are stiff with embroidery of style rather than embroidery of thought.

Can one imagine a greater pleasure to an inquiring boy of ten than, escaping from the severe rule of John Tiliander, to travel with his father, also a lover of nature, enjoying an unwonted excursion in spring through an interesting country? Even though the journey was to end in going to school, still school was a novelty; and to one with whom travel in his own country was a passion, this was a joy to stand out salient through life. The parting from home gave it the touch of pathos that thrilled the nerves and made the feelings more sensitive to every impression. They were off in the morning early, for they had before them a ride of over five-and-thirty miles. They rode, most likely, both on one horse. Good-bye, loved mother and fond little sisters! Their handkerchiefs are waved dry before the dear travellers are out of sight from the knoll on which they stand. Sweden in spring is one vast natural history collection, all careless of mankind, and Småland is the very pith and core of the conntry as regards entomology and botany, both of them our Carl's wild delights—objects that he loved as other boys love their boats and bats.

Småland is one huge moraine which has poured itself upon a lake, filling it, except where a few pools are left so mingled with the stones that it is hard to say where dry land begins and ends. It is said of Södermanland (Sudermania), which is likewise a confused mixture of lake and forest, that here the Creator omitted to separate the land from the water. This is still more aptly applicable to Småland, which is the superlative of this. Solitary it is, yet full of life that never allows the country to feel gloomy—the early heron fishing in the lake; the young trees springing all about; happy families of fir-trees, thick as grass—

> Those delicious self-sown firs, whispering,
> What has been shall be.

How vivid is the verdure of the spruce in spring, enhanced by that blue low distance to the northward, where the ground has been partially cleared; elsewhere, the hoary limbs of patriarchal trees harmonising with the primeval boulders, make all one grey mystery, into which the sharp eyes of our young Linnæus can pierce and he can find out treasures for his collections, where perhaps only a pair of kites are visible above the slanting splinter fences to a more ordinary observer— where some, perhaps, can merely see the gates which so often cross the Swedish roads to keep the cattle within bounds, and others see only the vast chaos of the land!

The ground somewhat changes character on approaching Alfvesta on Lake Salen, where the marsh marigolds, heralded forth by the cuckoo, as they say in Sweden, are flowering right out into the water beyond rushy peninsulas; and the ancient church of Aringsås is seen, with its curious detached belfry mounted on a wooden scaffolding, or, rather, a peculiar arrangement of wooden columns. Fergusson—who, however, is too hard upon Swedish architecture generally—says, 'The most pleasing objects in Sweden are the country churches with their tall wooden spires and detached belfries. If these do not possess much architectural beauty, they, at all events, are real purpose-like erections, expressing what they are intended for in the simplest manner, and with their accompaniments always making up a pleasing group.' Swedish architecture is mostly very simple, but fully expressive of its intention. There is no opulence or splendour in Sweden, not even in nature; the beauty there takes other characteristics—fair flowers, blithe birds and insects, and fair women. God manifests Himself in different ways in different countries, through other darkened glasses. In Sweden—its air, its snow, its social life, its moralities—all are pure; therein lies its charm. It is plain, but with the purity of snow. Nature and science go hand in hand in Scandinavia; art is left out of daily life altogether. Their school of painting is only nature transcribed, or set on canvas, with affectionate feeling but no ideal grace. There are some interesting runic stones at Aringsås.

Here Nils Linnæus stopped to dine at the house of

one of his wife's relatives—Professor Lars Johansson Humerus, a cousin of the Brodersons, and also connected with the Tilianders. The rector introduced his pretty boy, who evidently made a favourable impression upon the professor, for he counselled his father to send Carl to complete his education at Lund University, where he promised to be kind to him. The travellers then proceeded on their way to Wexio.

Their road skirted the lake, and, though still wooded and rocky, the scenery was less wild and unconquered by man than their own Stenbrohult, and the climate even milder than at Liatorp, the most fertile part of their own parish. Here the gooseberries were setting for fruit; at home they were still in blossom.

'All nature is alive, and seems to be gathering all her entomological hosts to eat you up,' says Sydney Smith of the wildernesses of Brazil. In Sweden, awakened nature at the close of May is arraying her entomological hosts for work or warfare. Still wood and rock; though some of the larger masses of stone here are not boulders, but the living rock, showing traces of glacier action in its rounded smoothness. These rock walls are stained red with microscopic lichen, against which the hoary stems of pine and fir, all powdered or dusty with their parasitic moss, and the coarser ragged lichen hanging about the low wiry twigs of the firs, look ruder and rougher than ever. The principal crop grown hereabout is telegraph posts.

This place is called Gemla; though why it bears a

name at all it is difficult at first to see, for only now and then a back settler's cottage is discoverable by the track that leads to it through the woods, and a few patches of pasture are carelessly enclosed here and there by the rudely made fences of splinters stuck obliquely between tall upright rods of irregular height. There are nowadays, however, several manufactories tucked away inside these rude forests, and served by the branch railway to Wexio. About four miles further lies Räppe, at the junction of the two lakes, Helgasjö and Bergquarasjö, on whose borders rises the picturesque ruined castle of Bergquara; and here the father and son are nearing the end of their journey. The ground is smoother; that is, there are fewer boulders than at Gemla, though occasionally a huge immovable block of granite stands right in the middle of a field of rye. They cross a wooden bridge over the narrowest arm of the smaller lake, and can see in the distance another large islanded lake, with a church spire and blue hills on its further shore, which lake is connected in the usual Swedish labyrinthine watery manner with the Helgasjö at Räppe.

This is the approach to Wexio, a clean, white, comely, empty-looking town, seated on a pretty blue lake which is part of a vast general water system in these parts, where there is now (in 1886) an esplanade with seats; and in the evening when the fashionable folks turn out to promenade it looks like any modern watering-place, only prettier and pleasanter than most.

It had not altogether this appearance in 1717, for the town has been stone-built since the fire of 1843; but in some parts its aspect is unchanged. The cathedral, dating from 1300, with its curiously battlemented tower and its six transepts, is one of the quaintest I ever saw. It focusses a pleasing scene as one sits in the leafy avenues of its close, admiring the truly Swedish mixture of its colours, red, white, and grey, set in foliage, and backed by a deep blue sky. Not until one comes to draw the building does one perceive the variety of its forms and tracery. It looks so simple with its six transepts all set in a row, yet it puzzles one's perspective more than many more seemingly elaborate churches.

A bridge under the railway leads to the—sea, I was going to write, the blue lake looks so like the sea from the avenues round the church; and a lofty bridge over the railway leads to the higher gardens they are laying out above the lake borders. Wexio is an attractive place, and none so small either; on closer view it has all the consequential appearance of a flourishing town; though, with a population of only 4,000, it looks over-housed.

It must have seemed a prodigiously fine place to young Linnæus, now seeing a town for the first time; but the happy hunting-grounds for natural history looked a long way off. Father and son repaired at once to the grammar school—now called the old grammar school, for a handsome new one has been lately built and the old school-house has been handed over to a lower class of boys, who, however, look very respectable and fairly well to do, besides being well-mannered—but this last is universal in Sweden. The old grammar school bears a family likeness to the cathedral, having a frontage of five gables in a row, and a large playing-field before it fringed with fine avenues of plane trees. An obelisk to H. Siögren, on a grassy mound, faces the school. It bears the motto *Aliis, non Sibi*. Lænerius, the rector of the school, was a botanist; the two rectors talked of botany and looked round the garden together. Now the father could jog home rejoicing: this brother botanist would be kind to his boy.

The Linnæi next went to call on Dr. Rothman, the successor of their relative John Lindelius as the physician in highest repute at Wexio, to bespeak his interest in the lad. While the elders were holding a long chat over botany, a subject of interest in common with both, we may suppose young Carl eagerly listened to their talk, and when it turned on medical topics—for it is more than probable that the rector took the opportunity of speaking with the doctor about his own ailments, as we know that he afterwards consulted Dr. Rothman on a malady to which he was subject—the observing Carl found objects of interest in the room to occupy his attention and a country child's natural weakness of wonder. They were pressed to stay to supper, but the hospitality was declined. Father and son spent their last evening together. They sat by the delicious lake talking in the sunlight, now at eight o'clock (May 28), glancing on the little rowboats out a-fishing:

the water so clear, the opposite shore reflections so soft, exquisitely tender—as the parent's heart. Young Carl, though he cared less for fish than for any branch of natural history, was yet interested in the unprofitable fishing of the men and boys from the turfy bank. The fish themselves had unsuccessful sport; it was equally amusing to see the numerous fish leaping up at the swarms of gnats above the lake, and the way the gnats darted away and escaped.

The Linnæi took their evening meal together, in a garden overlooking the lake (one can enjoy this even in Sweden on May 28), while watching the moon (at 8.20) rising white, dim, spectral, above the lake out of the mist—not silver, only just a dead white, gradually (at 8.45) becoming more normal in brightness.

The inhabitants of Wexio come out to wander up and down in the cool sweet air. The women affect fawn colour and rosy-pink, and brick-scarlet cottons over their ordinary grey-blue woollen clothes; these contrasts have a pleasing effect in the landscape. They wear white or pink kerchiefs on their heads; otherwise there is no especial costume in this part of Sweden. This kerchief is the national head-dress; it is worn of black silk or cashmere edged with lace on best occasions. All this town gaiety, which at another time would have been so brilliant and dazzling to the country-bred lad, loses its charm this evening while he clings to his father as the time of parting draws near.

Andres Celsius of Upsala had not yet invented his

thermometer, for he was a boy then—only six years older than our Carl; but the temperature on the following morning really stood at 13-14 (Celsius), making the fire in the white stove in the coffee-room of the inn very acceptable as, after stirring its embers with a brass trident, father and son sat down to breakfast off fish and eggs and a basketful of three shades of brown rye and barley bread, and the pretty maiden who waited on them brought some fine white bread besides as a special welcoming. '*Tak, tak*,' say they both—the Swedish for 'Thank you.' Since Charles XII.'s wars carried off the men women have always been waiters in Swedish inns.

The sun coming out makes it warm, but not yet oppressive, as they walk to the school by way of the sparkling blue lake. It seems higher water than last night, as if there were a tide in it. But feeling rises above the line of noticing such things as the moment of parting draws near. It comes, the embrace is over, the blessing left behind, and all the love that can be concentrated in one last yearning look, and the parent, with chill at his heart, foreboding on his mind, and a prayer at his lips, rides home through the pine woods, still tragic with their traces of snow-havoc. A young forest tree had been uprooted and hastily put in to fill a gap in the splinter hedge. Pastor Nils Linnæus looks at it painfully and then turns quickly away. How glorious were these forest aisles when solid with the crystals of winter, and when his boy played in them and lit all nature up for him with the gladness of his rosy face:

the trees a vista of white pyramids, each tree topped
with a pinnacle of snow! He rides wearily home
beneath the 'querulous fraternity of pines'; the light
dies out of the landscape for him now Carl is gone; the
boy is a child no longer; it is almost like losing him to
feel that the loved babe is gone, changed into another
form—improved, it may be, but the same no longer.
He is a son gone out in the world. Henceforward, the
pastor must enjoy his flowers alone; yet he will enjoy
Carl's interest in their growth in the rare periodical
visits home—rare of necessity because the family is poor
and journeys long—and Hope shines at the end of a
long vista of years. The pastor is not a great man,
though he had youthful aspirations; but Carl will be
great, and make a name that will be known throughout
the province—throughout Sweden, it may be. He may
even rise to be a bishop in the Church, for he is a lad of
noble promise. These duties of weaning oneself from a
parent's joys are painful—from the pang of shearing a
boy's golden locks to the greater grief of his severance
from the home world. Perhaps one feels these most
with the first and the last to go.

Wexio was a large world to Carl: the school and
gymnasium had in his time 210 scholars.[1] His progress
in the Latin school was not satisfactory—as to Latin;
yet everyone spoke well of his good conduct and pleasing
manners; but he was inattentive, and he took every
opportunity of escaping out into the country to collect

[1] Linnæus mentions this in his Skåne Journey.

plants. On holidays no pupil was so little found at home as Linnæus. How he admired the plants that everybody keeps in his window in Sweden! that is, everybody but schoolboys; it is difficult for them to have plants of their own. Dr. Rothman seems to have been kind to him—for his father's sake a little, and more for the look of bright intelligence which flushed his face when a word was spoken about botany; but a physician in full practice has not much time to spare over a boy of ten or twelve; if he 'tips' him or asks him now and then to tea with his children, it is the utmost he can do for him.

Carl had been a year at Wexio when his brother Samuel was born, in 1718. The parents had thus been absorbed in new joys and cares; a tragedy, too, had happened in the family; Abel Tiliander, rector of Pietteryd, was accidentally drowned in a well.[1]

In 1719 Carl was put under the private tuition of Gabriel Hök, who afterwards married his sister Anna Maria. This man possessed a milder disposition and much better talents for teaching than John Tiliander; but he could not overcome the distaste the boy had contracted towards the ordinary studies of the school.

Carl's progress was still slow. Not for three years did he receive promotion to a higher form in the school, called the 'circle';[2] and this time must be unusually prolonged or it would not be remarked upon. In the circle he had more liberty and leisure, and devoted both

[1] Diary. [2] Notes for Autobiography.

to the study of his choice. He wandered about the outskirts of the town seeking plants. Lænerius, the rector of the school, often talked with Dr. Rothman about the talent of the boy, and, being himself such a lover of botany, perhaps relaxed discipline in his favour. Stoever says he viewed his pursuits with complacency; at least, he considered them as innocent. 'He grew fond of a youth who so ardently entered into his own researches and displayed such extraordinary talent. He formed a proper judgment of his genius and application, while Carl's schoolfellows considered him as a vagabond who wasted his time in useless studies and running about.'[1] It is true he was not sent to school for that.

At sixteen he began forming a small library of botanical books, comprising Manson's 'Örta-bok,' or herbal, Tilland's 'Catalogue,' Palmberg's 'Serta Florea Suecana,' the 'Chloris Gothica' of Bromelius, and Rudbeck's 'Hortus Upsaliensis.' These latter he could not as yet understand, but he committed them to memory.[2] For this his schoolfellows nicknamed him the Little Botanist. Our nicknames in England are seldom so well-sounding, but boys are more mannerly in Sweden. He was confirmed in the cathedral of Wexio, by the bishop and another minister, in full lawn sleeves and copes of crimson velvet with great gold crosses on the back.[3] It is a pretty sight to see all the little fair heads of the girls as the deaconesses range

[1] Jardine. [2] Diary. [3] The Swedish clergy always wear ruffs.

them in their places in the cathedral before confirmation, the rosy boys hanging shyly about the columns of the church, waiting to be shown their places.

In 1724, when Carl was seventeen, he was removed to the upper school, or gymnasium, a separate building, where the higher branches of literature were taught. Here his tutors, like those of Newton at Cambridge, gave him up as a hopeless dunce. There was no modern side to a public school then; a lad had to fight his way against and through the classics. A test examination showed that his time and attention had been all absorbed in his eagerness after flowers and insects. His father was written to and a manual employment recommended. This was what the Swedes call a Job's post—a bad news-letter. The examiners were severe, and although in mathematics, and particularly in physics, Carl did well at Wexio, still the Greek and Latin grammars reigned supreme, and the tutors told him flatly he was fit for nothing but to be a cobbler.

His fate was otherwise decided. At one of his visits to Dr. Rothman he met with Tournefort's 'Elements of Botany.' Away went all remembrance of the examiners: from henceforth he could be nothing but a botanist. This was the keynote in his career.

Though a good and pious boy, he entertained an intense dislike of the study of divinity as a profession; his sense of duty to his parents fought against his dread of their forcing him into the ministry, for which he felt no vocation. He roamed the fields now more in distress of mind than for research: he wandered off as far as the royal tumulus of Amlech—Shakespere's 'Hamlet'; his disturbance of mind was, perhaps, equally great. Had Linnæus been able to read English he would have found a kinship in Hamlet's unwilling acceptance of life, with its problems not always adapted to man's varying mental constitution. To be or not to be a clergyman: that was the question. How dare assume to guide others, when every blade and leaf taught him his own ignorance? He could not receive the narrower doctrines —the only ones then current—as to what objects were the best worth seeking in this world. How was it, then, that his companions, who, he could not help seeing, were most of them less talented than he, were able without difficulty to pursue studies that for him were like beating his head against a stone wall? He was brought down to earth again. What! Were all his botanical excursions to be stopped?—his only pleasure at Wexio, where 'amid his wild-wood sights he lived alone. As if the poppy felt with him.'

Stoever proceeds in his inimitable way: 'Dogmatical acquirements, the Hebrew language, and the more solid branches of scholastic science had been forgotten amidst the allurements of the goddess Flora, and still continued to enjoy their usual share of oblivion.'

When we read passages like these our own ponderous Johnson feels less sesquipedalian. He seems light reading after seeing Stoever and other Germans disporting themselves like whales. Carlyle talks of how 'old German books, dull as stupidity itself—nay, superannuated stupidity—gain with labour the dreariest glimpses of unimportant extinct human things.' But they are always trusty as to dates, according to their light, and mostly as to facts, when one can get to the bottom of their meaning. Carl's parents took the complaints of the professors and lecturers of the college much to heart, foreseeing in the evil report the probable ruin of their fondest hopes. The mother argued thus: 'His father loved plants too, yet he got the divinity, or theology, into his head. Why could not Carl? Was all the rise in the family to go for nought? Was the peasant family to be again degraded to the ranks? He *must* toil at this uncongenial study for his forefathers' sake—and for theirs. His father had no money to give him, and the boy could not expect to live by picking flowers.'

His father came out to see him soon after this. The first evening was a happy one. The meeting could not fail to awaken pride and delight in his boy in his merely physical aspect. Nils saw himself young again, with added charms such as he could not remember in himself; besides, he had never been without influence over his son. Oh yes, all things would be set right by some mild yet firm parental talk. The lad had promised to be so clever in earlier days. But the next morning, after the masters of the school had pronounced him unfit for any learned profession, a cloud of sadness rose between father and son; they were no longer able to see each other's mind. The tutors had no opinion of Carl's abilities, and again counselled his father to put him to some mechanical trade—a tailor, or better still a shoemaker, a favourite craft in Sweden, and, I suppose, therefore the most profitable; it was, at all events, a secure livelihood.

The account in Linnæus's diary runs thus: '1726. The father came to Wexio, hoping to hear from the preceptors a very flattering account of his beloved son's progress in his studies and morals. But things happened quite otherwise; for, though everybody was willing to allow how unexceptionable his moral conduct was, yet, on the other hand, it was thought right to advise the father to put the youth as an apprentice to some tailor or shoemaker, or some other manual employment.' Good as is the evidence of the diary, it is only the rapid rough draft of the fuller and sometimes slightly differing account in the autobiography begun in Latin, and continued in Latin or Swedish by various hands from dictation, or compiled from conversations. This date of 1726 seems to be an error, as we know this event occurred three years before his admission to Lund University, where he went in 1727. With filial obedience Carl avowed his readiness to study divinity, but owned at the same time his want of inclination, his great aversion. His father therefore resolved to make his son 'take absolute leave of the muses'—old Stoever's expression—and to bind him apprentice to some honest cobbler.

Would Linnæus ever have sung at his cobbling shopboard like Hans Sachs? No, for Linnæus was no poet, no psalmist, no student of men. He would have reared himself a bower of greenstuff and followed with melancholy musings the movement of the flies in his window. He might have been himself lost. Who can tell? Would so strong a bias have created for itself an opening to the light if imprisoned in an uncongenial forced labour?

One person only appreciated the form of industry of the boy, of whom none spoke in any blame except that he had no taste for the grammar school routine—a thing not uncommon among idle boys. Yet Carl was not idle: there lay the problem. The rest never thought of solving it, only of smashing it open. This person was Dr. Rothman, the physician and medical professor in Wexio College.

The old clergyman, having for some weeks laboured under a complaint which perhaps had now been increased by his anxiety, was obliged to consult Dr. Rothman professionally, and, grieving at the seemingly wayward and careless disposition of his son, he opened his mind to the doctor, who kindly prescribed for both his mental and bodily sufferings.[1] 'Rothman intimated that he found himself equal to the cure of both complaints.'[2] The boy might arrive at eminence in medicine, as being more intimately connected with that branch of his own choosing. He counselled his not

[1] Sir W. Jardine.   [2] Diary.

being forced to the Church, but to a more congenial profession which would utilise his botanical studies. He finally offered to give young Linnæus board and instruction for a time if he were permitted to continue his studies at the gymnasium—not in divinity, but in medicine. At the end of a year they might see if a trade were really the better decision.

This was some comfort to carry home to the anxious mother.

'Life is not long,' says Dr. Johnson, 'and too much of it must not pass in an idle deliberation how it shall be spent.' But it is human nature all the world over to seize any delay in making a change for the worse something may turn up.

Rothman spoke kindly of the lad—of his diligence, his peculiar endowments for his favourite studies. The first praises of his boy sounded sweet in the father's ears. Rothman was himself an eminent man, celebrated throughout all—Wexio. No matter the area, the celebrity was the thing; he was first in Wexio.

Carl added his entreaties to Rothman's persuasion. Many times had he heard his father say that a young man ought to learn that for which he felt the greatest inclination, because the natural propensity of a person always advanced him most in point of perfection. He was right in a general sense. It requires the highest genius to fight its way through all drawbacks. It is like good roads and good walking-shoes to a traveller that the line of life should go with the general bias.

An actor or actress should not be overweighted by personal unfitness.

With a shrug of the shoulders at parental weakness, the masters who had urged his being articled to a shoemaker received back young Linnæus as one who was to fail in medicine likewise. They showed even less penetration than the easily-blinded father. Dr. Rothman had the clearest eyes of any of them. They gave it as their opinion that Carl was not endowed with such parts as would qualify him for any learned profession, grounding this judgment on the little progress Linnæus had made in Latin. No sooner, however, had Rothman directed him to read Pliny than his progress became rapid; because the contents of that author corresponded entirely with his own natural propensity. To this circumstance may be ascribed his predilection for Pliny and the laconism of his style. Yet he loved the Georgics even better.

The father had still to consult with his wife, who would be deeply hurt at the ruin of her hope of seeing her son a minister. Equally disappointing was it to the father, who had himself raised the family from the peasant station, to find it must return to the clay from whence it sprang—who had hoped to see himself surpassed in his boy. How should he break it to the mother, the proud ambitious mother, who was waiting at home listening for the splash of oars on Lake Möckeln for her husband's return with details of the lad's triumphs, that her boy was considered good for nothing?

Who knows, she thought in anger, the tutors themselves might be jealous of his gifts! Alas, poor fond proud mother, jaundiced now even to disbelief in Rothman! Her son would have to be a shoemaker after all! Oh, the sadness of that night! Vainly did Nils defend his own favourite pursuit. She who had loved flowers all her life now loathed them. Never should the babe Samuel have anything to do with natural history; he should not enter the hateful garden. The child should gather no wild flowers. This very restriction made Samuel in later years a botanist; but his love for plants not being so ardent as that of his elder brother, his parents were not deprived of the gratification of seeing him in due time become a minister. That is, his strength of purpose was not so great as Carl's, or his sense of duty stronger.

Christina, the daughter as well as the wife of a clergyman, felt more keenly on the point of family pride than Nils did. She felt the hope deferred that maketh the heart sick when her cherished wish had to be transferred to a younger boy. Carl's throne in her mind was vacant from henceforth.

Though Carl redeemed this suffering nobly afterwards, he was not morally so great as Banks. Our admirable English naturalist had the stronger character of the two, the wider mind, which can take to itself even uncongenial learning. As an instance, once when overwhelmed by his great love of flowers, he said to himself, 'It is surely more natural that I should be taught to know all these productions of nature in preference to Greek and Latin : but the latter is my father's command, and it is my duty to obey him. I will, however, make myself acquainted with all these different plants for my own pleasure and gratification.'[1] He immediately began to teach himself botany. Banks was a rich man's son, and might with more impunity than Linnæus have been idle.

I admit the faults of my immediate hero Linnæus; I have no wish to make him out perfect: he had many weaknesses. He was a great man for all that.

After much hesitation the parents at length consented to let their son follow the new line. How ardent became Carl's love of nature now! how happy his life henceforth till he was twenty! Linnæus entered with redoubled eagerness into his now encouraged studies; only disturbed by such hard facts as certain Swedish plants not being reducible to the rules of Tournefort's System, he could expand freely in the career he had hitherto pursued by secret and interrupted steps. The certainty and limitation of a settled plan of study concentrated his zeal and spirit.

Rothman gave his willing pupil instructions in physiology and botany, and pointed out, somewhat superfluously perhaps, the advantage of studying the latter science according to the system of Tournefort. Carl's lynx eyes had discovered the text-book before. He had already begun to arrange every plant in its

---

[1] The somewhat priggish sound of this is due to the sympathising biographer.

proper place, and even to doubt the situation of many species whose characters had not been properly ascertained. 'Rothman gave his pupil a private course of instruction in physiology on the Boerhaavian principles, that he might make more rapid progress. He was rewarded by his success.'[1] In both studies Carl made considerable proficiency.

Tournefort, however, gave him the first view of the conveniencies of arrangement and the beauty of system, and was doubtless the foundation-stone of his own later structure. In writing the life of an eminent man it is customary to speak first of his ancestors, of his parents being poor and honest, and so forth; his mental ancestry is of even more importance to his biographer. Linnæus's immediate ancestor, metaphysically, was Tournefort. His valuable book[2] was not only illustrated, but elucidated, by the insertion of a figure of a flower and a fruit of each genus. Carl saw nature by this fine strong light, as modern artists see the external movements of nature by the teaching of Ruskin. Little did Rothman think he was forming the mould of a greater botanist than Tournefort.

Tournefort, who was born in 1656, died in 1708—the very year after Linnæus was born—aged only fifty-two. 'He might have been alive now,' thought Carl regretfully as he turned over the book that was ablaze with light for him, 'and I would have walked barefoot[3]

---

[1] Pulteney.   [2] *Institutiones Rei Herbariæ*, Paris, 1700.
[3] Letter to Haller.

to sit at his feet had not that hideous accident destroyed him.'

Tournefort, weakened by his laborious travels in the East, was felled by a blow on the chest from the axle of a carriage—an injury from which he had not strength to recover. Tournefort was the first inventor of the Genera; therefore was he most immediately Linnæus's metaphysical father. To meet with him at the first unfolding of his mind was a regeneration to our Carl. Tournefort was his real tutor, then came Vaillant. Carl's own neatly kept little library consisted of books calculated rather to fire than to satisfy his curiosity. These works, he felt, were only the beginning of science; the fire laid, he longed to apply the match and fire the mass.[1] He attempted to arrange in systematic order the plants growing around him, which, being Swedish, varied considerably from the French examples of Tournefort and Vaillant.

He felt acutely the imperfection of even Tournefort's system. Oh, if he could perfect the system, or invent one which would be less incomplete! This was his boyish dream: a fine ambition for a youth of seventeen. Even then he began to feel the difficulties attending classification. He had already got beyond Rothman, who worked very contentedly with his favourite text-books.

Carl remained three years with the worthy Dr. Rothman, and gained his education. These three years at Wexio passed quickly, pleasantly, now that he had

[1] Jardine.

liberty of thought. It is not altogether surprising that at Wexio, although he lived there ten years, they hold little tradition of Linnæus. The old gymnasium, which now contains the Småländic museum, library, and collection of antiquities and coins, has since honoured itself with Linné's bust, but only one person here and there in Wexio knows that he studied here at all: for even in his day he was only known as an eccentric young fellow who wasted his time on things outside the school routine, causing some surprise as to why, considering his parents were poor, he was allowed to remain there at all. He might have wasted his time less expensively at home. He seldom shared in the schoolboys' sports; but the masters said, in more sarcastic but less witty words than Dr. Johnson used of himself, he contrived wonderfully well to be idle without them.

One Christmas Carl invited his more kindly preceptor Gabriel Hök to come home with him on a visit and tell him all about Lund University, where Hök was entered as a tutor. Here Gabriel saw Anna Maria, the eldest of Carl's three sisters, a pretty girl, if we may judge by her portrait, taken in later life, which now hangs at Hammarby. Hök, to please Anna Maria's parents, spoke well, indeed proudly, of Carl; all of which promoted the enjoyment of that pleasant Christmas holiday. Carl, it appears, did not return to Wexio, but stayed some months unsettled at home. Probably the parents feared to risk, or were unable to furnish funds for his entrance to Lund University, until his mother's relative, Professor Humerus, urged their sending him thither, and offered to provide maintenance for him so long as he should need it.

We can surmise how eagerly Carl accepted this offer by an entry in a pocket-book [1] of his of later date, where he says he flew to Wexio to ask from the rector the necessary testimonials for entering the university. He says he left his parents on May 1 (11), 1727, for Wexio, returning on May 2 (12) back to Stenbrohult with his testimonial.

On applying to Nils Krök, rector of the gymnasium in that year, for a testimonial for entering the university, he was given the following—a very curious example of the way professors then worded their certificates:

1727. 'Youth at school may be compared to shrubs in a garden, which will sometimes, though rarely, elude all the care of the gardeners; but, if transplanted into a different soil, may become fruitful trees. With this view, therefore, and no other, the bearer is sent to the university, where it is possible that he may meet with a climate propitious to his progress.' Signed Nils Krök, rector.

[1] This pocket-book, in the possession of the Linnæan Society, is an interleaved copy of *Operis agrostographici Idea, seu Graminum, Juncorum, Cyperorum, Cyperoidum, usque affinum methodus, authore Johanne Scheuchzero, M.D.* Tigurin. Acad. Nat. Cur. Philippo. Tiguri: Typis Bodmerianis CIƆIƆCCXIX. It is inscribed 'Carl Linnæus, Upsal, 1728.' It is interleaved throughout, and annotated in dainty hand-writing and carefully-drawn flowers. In some of the blank pages at the end is written, 'Vita Caroli Linnæi. Ens entium, miserere mei!'

Linnæus, who must have been amused at the arboricultural illustration, speaks of this as a not very creditable certificate.

He gives, in the pocket-book, his birth and parentage, and a list of his classes and masters at Wexio. The entry of farewell to his parents on his departure for Lund, August 14, 1727, seems to me to apply to his actually taking up his residence at the university after the long summer vacation. It is not likely that, after the hurried journey to Wexio in quest of the testimonial, he would have waited so long before entering himself at Lund as a student. Several circumstances in the story of his early days at Lund imply his entry previous to the summer vacation; the solemn farewell to his parents would have occurred only on his taking up his residence at the university.

When records are scanty one works best by putting likelihoods together, by following his road and describing what he saw.

CHAPTER III.

LUND UNIVERSITY.

> Buoyantly he went.
> Again his stooping forehead was besprent
> With dewdrops from the skirting ferns. Then wide
> Opened the great morass, shot every side
> With flashing water through and through; a-shine,
> Thick-steaming, all alive. Whose shape divine
> Quivered i' the farthest rainbow-vapour, glanced
> Athwart the flying herons? He advanced,
> But warily . . .
> Each footfall burst up in the marish floor
> A diamond jet: and if he stopped to pick
> Rose-lichen, or molest the leeches quick,
> And circling blood-worms, minnow, newt, or loach,
> A sudden pond would silently encroach
> This way and that.—*Sordello*, BROWNING.

IN 1727, when he was just twenty, Carl Linnæus was matriculated at the university of Lund, in Skåne, South Sweden, where his father had studied, and contended with poverty for some years, but where Carl possessed two relations who would be of great help to him in his studies. One of these, his cousin Carl Tiliander, **was a** student of some years' standing.

Speaking so well and so persuasively as **Carl did,** his mother still looked forward to his being one day a preacher. She hoped much from the university. Carl travelled southward alone this time: he was to meet his elder relative Humerus, who was a professor in the university of Lund, and who had promised to support him.

It was his birthday, May 23 (May 13, Old Style). What makes the date of Linnæus's birthday of moment is that nearly every journey of consequence that he took, and many of the chief events of his life, are dated from his birthday. It is true that this is just the time of the break into the pleasantness of spring, and therefore, naturally, the time to begin botanical excursions. Carl looked up fondly at the red cottage, his birth-place at Råshult, as he passed it. Poor as he might be, a garden like that would be sufficient for his happiness; surely he might hope to compass as much in life, or so little, as that. He walked on—he was to walk the distance of about eighty-four English miles in four days—carrying his knapsack, and resting at various farm-houses or priests' houses on the way. Twenty-one miles a day promised a pleasant walk. There is nothing more delightful to an active young man than a lightly equipped walking tour. Carl was lightly equipped enough, we may be sure, and he found much to entertain him on the way. Through life he always enjoyed travel even beyond the usual relish of youth.

A walking-tour is a more formidable thing in Sweden than elsewhere, when one reflects that Sweden, containing 170,000 square miles, is consequently nearly three times as large as England and Wales together.

Then the towns and cultivated lands bear so small a proportion to the fells, forests, and barren plains (superficially considered, a monotone of difficulty), which together comprise 3,123 Swedish square miles, leaving, when we have deducted 429 Swedish miles for the lakes, only an area of 247 square Swedish miles of meadow and cultivated land.[1] One may travel for miles without seeing a human being.

Carl walked lightly on with the brisk step of a youth who means to carve his own way to conquer the world. He had to depend upon himself now. He was happy, being filled with the great thoughts of what he meant to do, and with that longing to name and define things that marks the time of noontide in the mind. The sweet fanciful vaguenesses of childhood's dawn having vanished, with the dewdrops all about them dried, youth is the hour when one really possesses one's pleasures, instinctively realising happiness—

> Yes, as I walk, I behold, in a luminous large intuition.

His first day's journey took him through Småland with its shingle-roofed red houses and its red lichened rocks, with juniper underwood, above which wave the silver birches whisking the flowing streams lightly and airily as does the line of a fly-fisher. The land was fair, yet for nearly the first time in his life Carl's own thoughts occupied him more than did external nature. It is true he had made a long day of it: the butterflies had been asleep for hours. The white owl was

[1] H. Marryat.

blinking himself awake, the white ghost-moth, just emerged from his chrysalis, was trying his yet moist wing. At 9 P.M. the evening light reflected the banks and trees in a white lake to the eastward on his left; the western sky was still suffused with buff and pale pink when Carl entered Ousby, laden with specimens, and made for one of the wooden houses, raised on cyclopean stone foundations, where dwelt a brother clergyman of his father's, who received him with hospitality and a round lecture, such as he deemed good for youth, and for this youth in particular.

Off betimes, for Carl did not care for a second lecture. He crossed the Helga by a ferry—not across the river itself, but further down in the pretty islanded lake of Ousby. The landscape became more smiling and commonplace. But there is natural history everywhere. From the rising ground at Hastveda he could trace the great plain of Skåne below him. This ridge still looks like a devastated land, only peopled, apparently, by one long-legged stork with white body and black wings. This was of old the borderland between the Swede and the Dane: henceforward to the coast the people become more Danish.

Carl would soon descend upon another world, a world of level mediocrity—so it seemed to him as he looked down upon the reaches of distance without one salient point. Should he too soon be lost in that expanse, that waste? Dismal reflections in a boy are generally a sign of its being dinner-time, and no dinner, or not much, forthcoming. He ate his goat's-milk cheese—it was his breakfast—and his flat round rye biscuit, as large as our largest-sized dinner-plates, that he had slung round his shoulders by a string threaded through the hole in the centre. This is the Swedish bread of everyday life. He ate and felt better. His hand formed a cup at any stream. No, he was not lonely, his old friends were about him. The undulating ground here is still lined with whortleberry plants and polypodiums. One tasselled spruce above a rock reflected itself in the lake mirror; boulders were standing up in the shallow water; the lake was still surrounded by fir woods. Småland does not level into Skåne all at once in a hard sharp line: it melts away, blending two forms of beauty. The graceful white-flowered bird cherry,[1] as they translate the Swedish *hägg*, a favourite tree of Linnæus, and the aspen are still very common.

A lake with boats upon it, all setting southward, invited Carl to step into one of the fishing-craft and work his passage for about half a Swedish mile. The swallows flew dipping and curving by the low banks on the eastern shore. As they rowed away from these rocky slopes towards the west the signs of prosperity came thicker on; more linen webs were spread out to bleach, and boulders were cropping out among the corn—what was a sign of poverty to them was to him a token of wealth: he was more used to seeing the stones crop out of the whortleberry masses. Carl bought caraway biscuits

[1] *Prunus padus.*

and bread flavoured with anise-seed of a woman who was carrying her basket to the ferry, and the boatman offered him a drink of light beer from his firkin for his luck in catching fish.

It was still broad daylight, for all that he had lingered, when he arrived at Hessleholm, where he walked about the town, or rather village, with its neat wooden houses with steps up and down at the porch; houses not so dainty as in Switzerland, yet still pretty and inviting, set in gardens full of cherry blossom, which is in full bloom even so far south as Skåne at this date. Hessleholm is an increasing place now that the railway is made to carry off the stores of timber that its sawmills make available to the outside world. No lectures for Carl to-night: these good people are strangers, and he has already fascinated them by his silver tongue and all the wonders he can show them in their own surroundings. They do not allow him to go forth with dry rye biscuit: they force on him an abundant breakfast, and they pack up a neat dinner of white bread and rarer fresh meat, and tempting cream cheese, and pickled fish, and bits of angelica steeped for weeks in honey.

A third day's journey in the sweet fresh air—there is something intensely balmy about the air of Sweden in May—and a third day's pleasure. So this is Skåne, that he has heard so much of, as we in England hear of the mildness and fertility of Devon. One more ridge of limestone with a windmill on it, and now he is on lower ground, with meadows in the blue distance beyond the fringe of woods; another lake, the Tinga, with marshy borders and a little stone jetty built out into the deeper water, and rocky scenery to the left hand, on the east. Had it but a mild winter climate, how people would fight for this delightful land, thinks the passing stranger from the south.

Though in some parts the soil is still poor and heather-covered, there is an appreciable difference in its average value as compared with Småland. There are fine currants in blossom at Sösdala, and the church-spire at Mällby is set in what looks like amazingly rich land to Carl, and there are water-lilies in the meandering rivulet below the rude but pretty little double stone bridge. Here and there the land is fairly cleared for crops, yet it is often impossible to clear away the boulders so as to leave room to till the soil, notwithstanding that they gather up all they can into lines of rude stone fences. The stones are, after all, too many for the hands they have to lift them. 'Here they should plant woods to shelter the clearer plains,' thinks the young Linnæus, ever ready to set the world to rights. And there are woods hidden away behind the ridges of rock. There is one particular forest at Sösdala where the black stork builds her nest and hatches every year a brood of young ones, who disappear none knows where. That wheat is not yet sprouting here and barley is very backward, is what a traveller from the south would notice. But to Linnæus Skåne's vegetation seemed in advance of everything he knew, save that of Liatorp itself, his local fondness would reserve, and Wexio, which is mild as any part of Sweden. He already felt traveller enough to institute comparisons. What! pine and spruce-clad rocky hills again! From above it looked all one blue ocean of meadowland. He stands on another shelf of native gneiss rock, beyond the limits of the great moraine, which forms a shield to the fertile Scanian level. The boulders of the moraine are chiefly granite, the underlying native rock is either gneiss or limestone. The granite has a good smooth fracture which adapts it for walls, bridges, and the cyclopean foundations of the houses.

It is softer, prettier country now, with fine rich earth too, and pigs in plenty and brown cows, and beech as well as birch woods at Tjörnarp—a sign to Linnæus that he had come far south, for the beech was rare up in his country—and a foliage-fringed lake. The lakes run through Sweden like necklaces of pearls: no sooner is one rounded than another rises ready on the string.

Up hill and now down again among blue flowers. There is still much moorland scenery, with rugged wastes of heather and purling streams, and some unawakened water in still pools, and wheat just springing in the well-sheltered patches of cultivation. Up and down hill in reiterating succession, in long stretches of both sorts, for this landscape comprises half a province. Here it resembles some lowland Scotch or Yorkshire scenery, a wild sierra region with beech and birch woods intermingled with rushy swamps spangled with marsh-marigolds, the more elevated ground whitened with wood-anemones.

People are richer here—at least they seem so to Linnæus, who judges by the houses, built of great stones in cyclopean masonry, the fine pairs of horses browsing in the grass patches (I, in speaking to Southerners, dare not call them meadows), and oxen drawing huge stones on timber trucks. Carl now came in sight of the pretty Ringsjö, or Lake Ring, beech-fringed and beautiful, and the timber station at Hör, with the chips built up smoothly in large cone-shaped stacks. Here they would have hospitably received him for the night. A child winding blue yarn on a wheel by her cottage door smiled a welcome to the youth; it was very tempting, but he had planned to get on to the Bosjökloster. He took a draught of milk and trudged on. He soon reached the peninsula on which stands the Bosjökloster, once a monastery, as its name shows, and even in Linnæus' time ready to receive pilgrims, who used to come to it from far and near. Monasteries were then still numerous in these parts. Count Beckfries owns the Bosjökloster now, and pilgrims never go there, and tourists rarely. The famous oak tree, now forty feet in circumference, and the oldest tree in Sweden, was even then renowned; but it was less remarkable then than now that the best part of two centuries are added to its age. Next day would be the last of Carl's journey: next evening he would see Lund and be received into the arms of his Alma Mater.

He took the boat in the morning with the rest of the pilgrims, chiefly small traders, and rowed across the lake southward, leaving Stehag far off on the right hand. After landing the party the boatmen set to work with their fishing-nets and tackle. They had bunches of flowers tied to their masts; the country people had them tied to their staves and in their hats: nowadays they tie blossoming branches and bouquets to the railway-carriages, such is their fondness for flowers, their welcome to the spring.

The trees were bare here, the range of low hills looked purply-blue behind them. Linnæus was surprised to see Skåne's broader aspect so wintry. It was nothing like our usual idea of hot summer bursting upon Sweden all at once; this was certainly a slow-moving spring. What huge narcissus bouquets the people carry! and yet what shawls, and wraps, and thick frieze coats they wear! Larks and thrushes sang to welcome the abundant flowers, which were much more plentiful than leaves. On the hill-slopes everywhere were wild flowers in profusion—cowslips, orchis and marsh-marigolds, whose unfolding is the signal for the cuckoo to arrive and the roach to spawn. One field was blue with pansies; blackthorn blossom peeped out among the boulder fences; the birches were just dressed in tiny amber leaf, the cherry-blossom was in its first freshness, and the gardens at Eslof were masses of variegated and early flowers. It was a pleasant journey through varied pleasing country, presenting, besides the ordinary wooden houses and stone cottages with thatched roofs, and storks making themselves at home thereon, a view of several handsome country seats of gentlemen and nobles.

At last Carl really descends once and for all upon the plain of Skåne; in ten miles more he will be at Lund. He recognises the more fertile landscape of his father's description now, in a vast expanse of sunny green pasture sloping away downward into aerial grey, just marked by hedges, a few windmills, and pollard willows; and a nearer water-landscape of a still river, full of fish, half shaded by birch and alder not quite in leaf; and, beyond a foam of pear-blossom, a fine reach of blue level distance seaward.

Carl had turned aside from the road and now stood on the 'Saints' Hill,' from whence the view at sunset is so fine. Before him are the towers of Lund—Londinum Gothorum, the London of the Goths—superior to our own London in old, perhaps legendary times, when it had 200,000 inhabitants and we had less than Lund has now. Lund is situated on the small river Höjeå, which was formerly navigable for large vessels.

From this height Carl can see Malmö and the sea beyond; yes, and what is that fringe to the right, that range of further distant towers, melting in the horizon's gold? They are not trees, they *are* towers—the towers of Copenhagen. Now indeed he is a traveller; he sees another country! He must sit down to pause and gaze and think. That golden distance seems like his life spread out before him. He sits there at gaze, half dreaming, while the sun sinks, and then wakes with a start to remember he is a stranger to the town and has food and a night's lodging to seek. He is weary and somewhat footsore. It becomes chilly too, the sea breeze carrying the cold so uninterruptedly from the Tartar steppes causes him to shiver. It is colder here than further inland. The poplars are leafless, though budding, and hereabouts the young hedge leaves are quite pale green, almost white, as if grown in the dark. Still it is a prosperous easy-looking land, with slight undulations, and to all appearance well peopled.

Carl entered the town by way of the bishop's palace, the hospital and university buildings, through the grove of elms and horse-chestnuts round the cathedral. The name Lund signifies in Swedish a pleasant grove.

It is too late to present his credentials, such as they are, to-night; besides, the longer he can postpone the ignominious process of showing the Wexio certificate of his incompetence the better. He must sup and look for an inn, as he does not know the address of his cousin Carl Tiliander's lodgings; and as for his relative Humerus, the professor with whom he is to live, it will be too late an hour to present himself before a college don after he has shaken off the dust of travel and made the best of himself. Besides, a few more hours of liberty will not come amiss.

The Stadshuset Hotel is much too large and important for his purse. It is customary in Sweden to find the town-hall building, where there is mostly a large ball-room, used also for theatricals and meetings—the principal hotel. It was so in Wexio too, as Linnæus remembered. A smaller house, with its modest ' *Rum för Resande*,' with supper, was soon found. It had been light enough to write long after nine o'clock, but the excited Carl found so much to record that he lighted his candle and sat up late to finish the journal of his travels.[1] Through life he kept a careful diary, not so much of personal occurrences as of his observations. He slept like a fossil.

Carl had told the people to call him early. He had no watch—a possession well-nigh indispensable in Sweden, where you never know in summer when to get up, nor when to go to bed, for daylight is no clue: in winter it is worse, for the darkness is then as perplexing as the overbright daylight in May and June. But it seemed, from the many noises of a town, to be nine or ten when he awoke. They must have forgotten to call him. Land of the great Gustavus, could it be that Swedes could thus forget a guest and break their promises? At length they knocked. They called it six. 'I don't believe it,' muttered Carl; 'the people are too wide awake in the streets for Whitsun week.' People do not keep such outrageously early hours in Sweden as in Germany: the daylight keeps them up later at night.

The market was being held in the open place before

---

[1] This one, alas, exists no longer.

the Stadshuset, and it was raining—a small steady rain. Carl did not wait to breakfast: he expected soon to find his cousin Carl Tiliander. A procession of the students was marching across the market-place, with a military band, and rabble following. Carl looked among the students for the other Carl, but could not distinguish him; it is true he might not have known him. He must go and inquire for him at the Akademiska Förening, and also for Professor Humerus.

On his way thither he looked up at the fine white Norman cathedral. It is really a grand building. To Carl it seemed stupendous, with its vast portal and lofty granite towers. It was suited to the time when Lund really housed 80,000 people, now dwindled to 12,000. When desolated by Charles XII.'s wars the town had only 680 inhabitants. How much murder one man may do and not be hanged for it! Carl entered the church, the doors being open, which is not usually the case out of service hours in Swedish churches. A funeral was going on; not actually the service, but the bier was lying at the foot of the seventeen steps leading from the nave to the transept, from whence two more lead to the choir, and again three steps to the high altar. It was evidently a person of consideration who had died, for the coffin was covered with wreaths, flags and memorials, and several persons stood watching the bier. Awed, but not much interested, Carl walked round the church, whose gilt and coloured roof was then only a shadow of its present self, for it has been carefully restored. He walked up on the right-hand side, where eight steps lead to the raised chancel, examining the monuments, placed tablet-wise in niches, of an architect and a king and queen.. These look early in date, though, the inscriptions having been tampered with, one cannot be precise about it. Carl was afraid of being shut in the church, so he hurried past the old carved-wood stalls round the choir, by the bay where the model stands of the building as originally planned, and through the archway facing this bay, where niched winged figures stand on grotesque animals, which have a Byzantine look, doubly strange in Sweden. These sculptures still bear traces of their former colours. The great square pillars of the nave, and the great rounded pilasters with their chamfered capitals, are as imposing as the best Norman work in France. Carl did not then enter the mighty crypt, lighted by ten windows and supported by twenty-four pillars—'the most beautiful and majestic part of the church'—which forms a ground-floor storey to the high-raised chancel: all of which reminds one of St. Denis, near Paris. Here are the colossal images of the giant Finn and his wife, said by legend to have built the church. These are huge stone figures clasping the great columns, which also have chamfered or sculptured capitals. The story read to Carl, later, when he had time to think about it, as if the old Scandinavian pagan heroes were buried in this crypt on the establishment of Christianity: or as if in 1080 the powers that then were in Scandinavia built this church under the direction of architects from France.

This cathedral is said to have been consecrated by Archbishop Eskil, an Englishman, in 1145. It is pure Romanesque or Norman in its style, and in its sharp-edged whiteness reminds one much of the Conqueror's and Matilda's churches at Caen. When this was built Lund was styled the capital of Denmark. It was often the residence of the Scandinavian kings.

Fergusson, who is never enthusiastic about Swedish architecture, says: 'The cathedral at Lund is older and better than either of these (Upsala or Linköping).[1] It was commenced, apparently, about 1080, considerably advanced in 1150, and the erection of the apse must be placed between these two dates. The little gables over the apsidal gallery seem part of the original design, and are the only examples of the class we possess. With these the whole makes up a very pleasing composition.' I wonder at the usually perceptive Fergusson not recognising above the fine exterior arcade the gabled corona, typical of the crown of thorns, for this meaning is well known to even ordinary writers on Swedish architecture. It is not the only example of this in Sweden, and a church in Gothland has the same gabled corona. I do not delight in gush, but one may express feeling, and Fergusson is really too calm. Its contrast with the Swedish wildernesses makes Lund Cathedral

---

[1] In Fergusson, the name Lidköping is manifestly a misprint for Linköping.

doubly impressive as a stately relic of the dawn of Christianity in Sweden.

Leaving the cathedral, of which one watcher by the coffin was the only living tenant, Carl hastened through the elm groves on his way to the university. In his hurry he did not perceive the approach of a student who was diligently absorbed in a book. They jostled each other, and Linnæus recognised Carl Tiliander. To see his cousin and to claim his friendship was one action with our Carl; but Tiliander was cool and did not respond to Linnæus's overtures. The sight of the unfortunate certificate was sufficient to make the rising young student, who was one day to be a professor in the university, pause before he proclaimed his kindred with one who seemed at best an unpromising young scamp. He would not help him other than by reading him a lecture for his good, and Carl never relished such.

We are nowhere told what was Carl Tiliander's relationship to the John Tiliander who was Linnæus's early tutor, but we may be quite sure that whatever Carl had heard from John about the boy was bad. This Carl was an eminently respectable youth—a bit of a Pharisee, I fear. He was not, as has been supposed, a professor at Lund on Linnæus's arrival in 1727; he was then only a distinguished student: he became adjunct teacher in Philosophy two years later—in 1729. This Carl was a celebrated man in his family; he was rector of Jönköping in 1741 and later a Doctor of Divinity. He was twice delegated as representative to the Swedish Diet. He coldly advised Linnæus to do the best he could with his awkward certificate, lifted his college cap, and passed on. The bells were clanging loudly for the funeral.

Indignant and astounded, our Carl stood rooted to the spot; never, if he starved first, would he ask a favour of a Tiliander who could thus heartlessly disown him. Would Humerus do the same? He almost dreaded now to meet his relative the professor, even though he had expressed himself in terms so kindly. The rain fell faster than ever. On leaving the shelter of the large horse-chestnut trees Carl passed the open square, now dignified with the statue of the poet Tegnèr, towards the red-brick round-arched building of the Akademiska Förening. Here the funeral procession was mustering to move towards the cathedral. The white-capped students, assembled under umbrellas, were following a grand display of banners with black cockades. The flagstaff of the building was twined with black. Linnæus waited while the procession filed slowly by at the foot of a mound with three rough stones set upright, surrounded by four rude slabs—a runic monument—and asked a bystander whose funeral it was that was thus honoured.

'It is that of a professor in the university—Professor Humerus.'

Linnæus staggered backward, but recovered himself, and following the procession to the church door, entered, and looked again upon the coffin of his only friend in Lund. It seemed he was chief mourner there. What was he to do? He went out of the cathedral with the others and still followed the procession, which now bore the coffin beneath the banners, the chaplets and mementoes being carried by the principal students, Carl Tiliander walking among the first. They carried the coffin first to the Kloster church (near the present railway-station), where an office was recited, and conveyed it, now on a funeral car, to the cemetery on the high ground to the east of the town. White-capped students carried the banners, professors and students of the highest grade came next, the whole body of the students following to solemn music of a martial kind.

What was Linnæus to do now? He must after all bind himself apprentice to one of the numerous shoemakers in Lund. Skåne abounds in shoemakers, for all that many little boys run barefoot. That trade is overcrowded, for here, as in Denmark, it rains shoemakers and shoemakers' boys.[1]

They were all departing, when one of the principal men forming the procession perceived Linnæus, and struck by his appearance of dejection as he sat himself despondently on a tombstone near the late professor's grave, he came up and spoke to him. It was Gabriel Hök, the suitor of his sister Anna Maria. Hök recognised him at once.

'Hallo, Carl! what are you doing here?' or its equivalent in Swedish.

[1] Danish proverb.

Hök deeply sympathised with Carl's misfortune in finding his relative and protector dead on his arrival. He looked at the unflattering certificate from Krök of the gymnasium at Wexio, and decided he had better not hand it in. The case was urgent. Hök took the responsibility upon himself, and used his interest to procure Carl's admittance into the university, and, withholding the doubtful testimonial altogether, introduced him to the dean and rector as his private pupil and procured his matriculation.[1] Thus, by Hök rather than by Krök, Carl's name was enrolled in the classes and the injurious document suppressed. He underwent with credit the matriculation examination of the dean and of Papke, the professor of Eloquence. He always had a silver tongue; if he spoke he prevailed.[2] Having thus settled this important matter, Linnæus was enabled to pass the vacation in peace at home; and, perhaps, with Hök's assistance, prepare for his first term. We are not told how Linnæus found means to attend the lectures of Kilian Stobæus, the professor of Botany and Medicine, which he mentions as beginning on August 21, as he had no money to pay the fees; but he did attend them, and these lectures enriched and rendered more exact the scientific knowledge of our young botanist.[3]

---

[1] Diary.

[2] Papke's examination is said to have taken place in August 1727, which has caused Sir J. E. Smith to suppose the matriculation was in August. Better evidence goes to show these events took place in the spring.

[3] Stoever.

His attention and diligence interested the professor, who pointed out to him the means of making a *hortus siccus*. Linnæus at once began drying plants and glueing them on paper. The dry air of Sweden is favourable to the drying of plants. Linnæus always dried his plants and fixed them with isinglass, each on a half-sheet of paper. I dare say it was through the friendly offices of Hök, himself at this time a poor man, that Stobæus was apprised of the ardent student's indigent condition; so that Linnæus found in his extremity of need a second good physician ready to hold out a helping hand to a struggling young brother. Like the kind Rothman of Wexio, Stobæus offered him accommodation free of all expense in his own family, and here Carl for the first time in his life met with a well-arranged collection of natural history.

This fact of his being again gratuitously received into a family proves Linnæus's good behaviour and manners, for we never hear of the ladies of these families objecting to him in any way. Stobæus had very bad health; he was one-eyed, besides, and lame in one foot. But what nature had denied him in bodily advantages was amply compensated for in the excellence of his disposition and the superiority of his mental attainments.[1] This was a delightful life. Carl's mind grew apace. He became acquainted with curiosities he had never seen before. The Natural History Museum of Lund contained a fine collection of birds and snow-white

[1] Stoever.

squirrels and winter-clad foxes from Lapland, besides minerals, shells, plants, birds, and other creatures, each one a specimen of a vast family out in the wide world. The present botanical garden of Lund did not then exist. The botanical garden of Carl's time flourished upon what is now a waste space in the form of a neglected shrubbery, where a few ancient cypresses with gnarled stems, old enough to have known Linnæus, grow at the back of the old university building [1] where Linnæus studied. This is an oblong brick building of the Renaissance mingled with a bastard Romanesque, in three storeys, with quadrangular turrets at the angles and a rounded tower in the centre, loftier by a cornice and an additional storey than the main building. This central tower has a pure Romanesque portal by which a winding staircase leads to the library [2] and reading-room. The books and pamphlets are arranged in open frames reaching to the roof. The grove of horse-chestnuts in front of this building must have been respectable young trees in Linnæus's time.

A small red building close by, led up to by a flight of steps, also near the large red-brick mansion of the Akademiska Förening, is the Kultur Historiska, one of the most interesting spots in Lund to Linnæus, though it might now be overlooked among the more elegant white stone buildings of the new university, standing on a terraced pedestal of granite, in Vitruvian Classical style, with pediments and sphynges above the cornice of

[1] Curia Lundensis.   [2] Universitets Bibliotekets.

the central hall. REGIA ACADEMIA CAROLINA is inscribed on the garden front of this new university, which has been built within the last six years. At the entrance door are four colossal female figures with tablets inscribed Theologia, Juris-Scientia, Medicina, Philosophia. This assemblage of new and old buildings gives a grace and dignity to Lund. PALÆSTRA ET ODEVM sufficiently designates the intention of another brick building completing the group. Tegnèr's statue faces them all. The present botanic garden is on the eastern outskirt of the town. There are some large tree-ferns in the hothouses, and the garden is a fine one, with borders of poet's narcissus (in May) a foot deep in long continuous chains; but, excepting these, there is a better display of flowers in the windows of the streets leading to the gardens. A notice was posted up inviting the students on May 28 to go on a botanical excursion to Refta and Fogelsång, this latter a favourite haunt of Linnæus.

They know nothing of Linnæus now at Lund, but they are very proud of their poet Tegnèr and his house in the Klostergade. The students do not learn modern languages; Greek, it appears, they speak fluently; a little more German would be more convenient, and perhaps English; French they do not aim at. Their manners are more Gothic here than in the rest of Sweden, from their proximity to Denmark, where people are less polite, though a great deal of capping and bowing goes on. But to return to Linnæus.

He was allowed to attend Stobæus's demonstrations

of shells, petrifactions, and molluscs, which were exhibited to Matthias Benzelstierna and Retzius, two private pupils of Dr. Stobæus.[1]

Plants remained, above all, his favourite study. His botanical arrangements so far were made entirely according to the system of Tournefort. His experimental knowledge, drawn from nature, was rendered regular, exact, and more extensive by that obtained from books.

There was also a young German student, Koulas by name, who lived with Stobæus, and to whom, among other indulgences, was shown that of having access to the Doctor's library. Linnæus formed a close friendship with this young man, and in return for teaching him the principles of physiology, which he had learned from Dr. Rothman, he obtained books by means of Koulas from Stobæus's library, which contained the most valuable works on botany. Linnæus's candle was often seen burning far into the night, to the terror of Stobæus's mother, who was very old and a bad sleeper. She desired her son to chide the young Smålander for his carelessness.[2]

Carl's candle was inimically observed by another person, a student named Rosen, higher in the university than himself, and a friend of Carl Tiliander, as well as a pupil of Stobæus. This young man, Nicholas Rosen, who had been till now Stobæus's favourite pupil, was jealous of the favour shown to the young Linnæus at times even over himself. Carl was so eager, so clever

[1] Diary. [2] Ibid.

and original in his observations, that it is no wonder a man like Stobæus enjoyed having the youthful zeal and brilliancy about him, encouraging his own drier studies and reinvesting them with the poetry they might have forgotten.

Now was Rosen's opportunity. We may admit that he really thought badly of his new rival from what he had gathered from Tiliander, who was also honestly entitled to his opinion; he thought it would be well if the professor's eyes were opened to the fact that he was wasting kindness on a worthless subject. He persuaded himself he could not bear to see the good professor deceived; for that Linnæus would disappoint him Rosen felt sure. Those two model young men, Rosen and Tiliander, were never without excellent motives.

'Do you see, sir, that light in Linnæus's room? He always keeps it burning very late.'

'Pooh! it is nothing.' Stobæus's second thought was, 'I fear the poor boy may feel ill.'

Rosen sneered politely.

'It may be so, but he loves company, and people passing his door have fancied they heard the sound of cards.'

The Rosen doubts crept into the cockles of even the professor's unsuspicious mind when night after night the lamp shone on the trees outside. What a pity if that nice, clever fellow should be tempted into practising what were then called the lighter vices! He was known to be of a social, convivial turn, and fond of company.

He might be making merry with the servants while the family had retired to rest. Come what would, the good Stobæus resolved that at all cost of unpleasantness to himself the boy should be saved. He burst into his room at eleven o'clock, and there sat Linnæus intrenched with the works of Cæsalpinus, Bauphius, Tournefort, and others.[1] These were his companions. Stobæus ordered him at once to bed after making him confess he had persuaded Koulas, the German student, to take the books out for him; but, delighted to find his favourite reinstated in his good opinion, he gave him free access to his library and made him one of the family, treating him, in fact, like a son.[2]

Professor Hök was always kind to Carl, but his having taken to the medical branch of study drew him out of Hök's supervision, he being a teacher of Divinity.

Carl had his livelier pleasures too—the students' carnival of Valborg's mass eve,[3] the *Walpurgisnacht*, when they light the Valborg fires. They collect materials for a bonfire on the highest and nearest hill, and the young people go up and fire the beacon and dance round the blaze in a ring, and tell fortunes by the flight of the storks. There were likewise the Midsummer festivities, with fireworks and dancing. Carl also was of great assistance to his protector in his profession. Stobæus being perpetually harassed with applications for medical advice from the nobility of Skåne, Linnæus was sometimes called to write letters and give advice in

[1] Stoever.   [2] Diary.   [3] Valborg's Day is May 1.

the Doctor's stead; but when he wrote a bad hand he was usually sent away again.

Besides keeping his regular herbal, Carl made excursions into all the neighbouring districts, exploring the animal as well as the vegetable kingdoms of nature. In an excursion to 'Fogelsång, in the spring of 1728, with a brother botanist, Matthias Benzelstierna, Carl was attacked by an accident or malady—for it seems uncertain which it should be called—common to the inhabitants of the Baltic and Bothnian coasts. A small animal is said to penetrate the skin and bury itself so deeply in the flesh that it leaves only a black dot at the spot where it entered. Unless immediately extracted, the effect of the animal's poison is to cause inflammation and gangrene with great rapidity, and death in the course of a day or two, or sometimes within a few hours. That this malady is indeed caused by an animal has been doubted and denied by scientific men; but Linnæus was convinced of its being so, and notwithstanding the suffering he endured while a parish priest was kindly acting as surgeon and extracting the substance, about half-an-inch in length, from his arm, he carefully examined it, and in spite of its injured appearance, pronounced it to be a true *vermes* and called it the *Furia infernalis*, from an idea that it realised the description of the fatal powers ascribed by the ancients to an imaginary animal so named.'[1] Linnæus utilised his mythological and other classical studies as aids in

[1] Smith.

the nomenclature of his discoveries. He was at this time especially interested in examining the lower forms of animal life.

Most Swedes think this furia is no worm, but that it owes its origin to a poisonous matter injected into the flesh by the sting of an insect. Though fruitless the result of all the researches made since Linnæus's time to discover an example of this worm, yet the disorder is common in the fenny parts of Eastern Sweden in autumn.[1]

Darwin, in his book on worms, says, 'In Scandinavia there are eight species, according to Eisen, but two of these rarely burrow in the ground, and one inhabits very wet places, or even lives under the water.' It was most probably a moist place where Linnæus was botanising; but Eisen says nothing about stinging worms, and Darwin does not concern himself with flesh-burrowers. In Scandinavia worm-burrows (in the earth) run down to a depth of from seven to eight feet.

[1] Linnæus thus describes the *Furia* in his *Systema Naturæ*: 'Habitat in Bothniæ Sueciæ Septentrionalis vastis paludibus cæspitosis; ex æthere decidua sæpe in corpora hominum animaliumque momento citus penetrat summo omnium dolore, immo interdum intra quadrantem horæ præ dolore occidit, quo et ipse Lundini 1728 laboravi. Anima! nonnisi rude siccatum vidi. Animalibus chaoticis videtur proprietatibus affine. Quomodo aera petat, unde decidit a solstitio æstivali in hyemale, nullus dixit.' Linnæus was no deep classical scholar: his Latin was fluent rather than accurate.

[2] Sir J. E. Smith. Linnæus's pupil Solander has recorded several cases of this accident or disease, and describes the animal as if he had seen it, in the *Nova Acta Upsaliensia*, vol. i. p. 55. The *Furia infernalis* seems an animalcule one-sixth of an inch long. Dr. Solander describes it as dropping out of the air in autumn. Art. 'Furia' in Rees' *Cyclopædia*.

Linnæus appears to have been seriously ill on this occasion, as both his biographers [1] remark that the skill of Stobæus saved his life. His own diary says differently: 'The arm immediately became so swollen and inflamed that his life was endangered, especially as, Stobæus being about to set off for the mineral waters of Helsingborg, he was left to the care of ———. Snell, however, having made an incision the whole length of his arm, restored him to his former health.' [2]

Linnæus had lived with Stobæus about a year, and the professor gave him hopes of becoming his heir, as he had no children.[3] But now, in order to recover his health, Carl went to pass the summer vacation with his parents in Småland, and here he met his first friend, Dr. Rothman; it is very probable he went to Wexio to see him, and the doctor advised him to leave Lund for Upsala, as a superior school for medicine and botany. Linnæus, too, greatly desired to see more of the world and widen his learning, and he resolved to go to Upsala. How to compass it was another matter.

His mother sighed to see Carl employ his whole time in glueing plants on paper, to the delight of little Samuel, who also loved plants better than Latin, and at last she abandoned her long-cherished hope of seeing Carl become a preacher. Linnæus's young mother had been passionately fond of flowers, and was always melancholy from the frosts of October until spring; yet she now

[1] Stoever and Pulteney.　　[2] Diary.
[3] Ibid.

solemnly adjured Samuel to look upon all flowers as prickly thorns and stinging nettles.

'But what is Carl to live on?' she asked.

'Never fear, mother, I will work my way'; and Rothman said the same. They all believed in him who believed in himself.

'When I was as you are now, towering in confidence of twenty-one, little did I suspect that I should be at fifty-four as I now am,' thought the father, unable to supply his first-born son with the necessaries of student life; perhaps he only felt what Dr. Johnson put thus into words. Rothman hinted the possibility of Carl's talents gaining for him a pension from Government that his studies might be utilised for his country, and the great likelihood of one of the many royal and other foundations of Upsala falling to his share. The hint lighted the spark of hope, the hope at once became a conviction in Carl's breast, and with a light heart, light luggage, his parent's blessing, and 200 silver dollars—reckoned at about 8*l.* sterling, his whole fortune—all that his father could spare him, or his mother save—he set out for Upsala to make his path to fortune and to fame.[1]

Linnæus was a self-made man. It is as a man, and in the history of his self-making, that he is more interesting to this generation than as a scientist. He left Wexio, Lund, and even Upsala, with a

[1] A. de A. Fée says it was 100 crowns. Were these écus, or were they kronor? 100 kronor would be little over 5*l.*

reputation utterly disproportioned to his great abilities. He had not consulted Stobæus about his removal to Upsala, although he must have written to inform the authorities at Lund of his intention, and asked for a testimonial of his attainments, as this time he carried a splendid Latin official testimonial from the rector of the university, in which he was called 'politissimus ornatissimusque dominus,' and was declared 'to have conducted himself with no less diligence than correctness, so as to gain the affection of all who knew him.'[1] This testimonial, addressed to the 'Candido Lectori,' is signed Arvid Moller, rector.

'With the stillest face, more touching than if it had been all beteared,' the still-young mother watched her boy depart, the stalwart son, losing with him her certainty of finding a protector to herself and her little children, the youngest girl a mere infant. Father and mother then turned and again sobbed in each other's arms and prayed for their darling, the hope of their age and weakness, for whom they had no other help than prayer. It was answered openly.

[1] Smith.

CHAPTER IV.

UPSALA.

In the very beginnings of science, the parsons, who managed things then,
Being handy with hammer and chisel, made gods in the likeness of men;
Till Commerce arose, and at length some men of exceptional power
Supplanted both demons and gods by the atoms, which last to this hour.
Yet they did not abolish the gods, but they sent them well out of the way,
With the rarest of nectar to drink, in blue fields of nothing to sway.
           J. C. MAXWELL.

UPSALA is distant from Lund seventy-five Swedish, or about five hundred English miles; from Stenbrohult it is eighty-four miles less. No biographer tells us how Carl made the journey, whether by sea or land, and those who mention it loosely give Michaelmas as the date.

His previously mentioned pocket-book [1] says Carl took his departure on August 23, 1728, arriving at Upsala on September 5. It names Ekesio, Skenninge, Örebro, Arboga, Köping, Westerås, Enköping as his route. The writer carelessly inverts the three last names, which if taken in that sequence would lead him

---

[1] Belonging to the Linnæan Society.

directly away from Upsala. Most travellers, poor, hurried, and unencumbered as he was, would have selected the more direct route by Jönköping up the Vettern and Hjälmar lakes, whence a short road across country would bring them to the Mälar, giving direct water communication to the very quays of Upsala. I can only account for Carl's choosing the longer and more expensive route by his considering the land journey would afford him better opportunities for study on the road.[1] I do not purpose describing this line of country, because we shall travel over all the region of the Vettern with him in his later and less hurried tours. Carl's journey was of necessity hurried, for, having only 8l. sterling, representing his whole patrimony, to embark on life with, he could not delay nor turn aside to visit objects or places of interest.

He arrived at Upsala, perhaps the poorest student who ever entered her walls.[2] I do not deny it, but the authorities have contradictory ways of making it out. Stoever says 'he had 200 silver dollars, worth about 8l. sterling.' This might be so if they were *German silver* dollars, but 200 thalers would be 30l. 200 francs would be nearer the mark, or Swedish kronor not so very far off, but the Swedes did not reckon by kronor in those days. I dare not contradict Stoever, lest I should rue discovering inaccuracy in a German; and

---

[1] His journey averaged thirty-two miles on each of the thirteen days.

[2] Stoever.

every little duodecimo biographer of early in this century has followed Stoever, *via* Pulteney, securely, and said Linnæus had 8*l.* One writer kindly allows him this sum annually. 200 silver dollars make about 40*l.*, varying of course with the dollar you reckon by. There is a considerable difference between 8*l.* and 40*l.* to a young man in any country; and 40*l.*, according to the value of money in Sweden at that time, seems a good deal for Carl's father to have spared. Money seems to be measured in Sweden something on the Scotch plan of punds instead of pounds, and the cost of living is still small at the Swedish universities. Lund has now about 600 students, Upsala double that number. 3*l.* a month at Lund and 4*l.* at Upsala will cover all the student's expenses. If this be so now—and it is an admitted fact—Carl's 200 silver dollars made a fair first year's allowance for him, even deducting a small sum for his journey.[1]

---

[1] Before 1777 accounts were kept in dahler of 4 marck, or 32 öre, either in silver or copper coins; the former being reckoned at three times the value of the same denominations of the latter. By the regulations of 1777 (which was the reckoning used when Stoever wrote his history) the specie riksdaler was to pass for the same value that 6 silver dahler or 18 koppar dahler formerly did.—Kelly's *Universal Cambist*, 1835.

The single ducats (the common gold coinage of Sweden) were to pass for 1 riksdaler 46 skilling specie; or 11 dahler 24 öre silver; or 35 dahler 8 öre copper.

The silver dollars used in Linnæus's youth were coins of Frederic and Ulrica Leonora, showing the two sovereigns side by side on the obverse, the reverse the three crowns of the realm; and the rarer pieces of Charles XII. with the crossed arrows, a crown, and a star on the reverse. The rapid changes in value of the coinage after Charles XII.'s wars causes the difficulty in reckoning Linnæus's funds.

The Enköping road, by which Carl entered the town, leads down the hill directly through the group of university buildings. The ground-plan of Upsala looks imposing on the map; but as all the 'new town' is as yet unbuilt, we see it pretty much as Carl Linnæus saw it on the day he entered Upsala.

Was it an omen that the first person he knew by sight was Rosen, his antagonist at Lund? The youths met coldly and soon parted. Rosen and Linnæus were about as unsociable as Swedish milestones. Carl gazed with more interest on the town itself; but, neglecting the fine cathedral, he flew to the botanical garden—not then what it is now, and vastly different to what Linnæus himself made it—and he thought it insignificant.

The Botanic Garden now has well laid out walks and alleys, tall screens of clipt limes and lower hedges of hornbeam and other close-grown greenery sheltering the various gardens, and a fine botanical lecture-room, built in classic style with a peristyle, within which the object that first attracts the visitor is the marble statue of Linnæus by Byström. The professor of Botany resides near the entrance to the garden. The present garden is on the high but sheltered ground behind the castle; the botanic garden that Linnæus saw was on the level ground on the opposite side of the river.

Disappointed in the garden, Carl, impetuous in all his ways, flew up what is now the Carolina Park, and away by its steep alleys to the hill whereon the castle stands.

He made himself master of the bearings of the town, and with experienced glance at once fixed upon the best site for a botanic garden, when he, the radical reformer, should once get a voice in the matter; he examined the place with curiosity, considering what improvements he should make when high in the university—for a great man he determined and fully expected to be. On the whole, he was pleased with the view of Upsala. These are his own words: ' Upsala is the ancient seat of government. Its palace was destroyed by fire in 1702. With respect to situation and variety of prospects, scarcely any city can be compared with this. For the distance of a quarter of a Swedish mile it is surrounded with fertile corn-fields, which are bounded by hills, and the view is terminated by spacious forests.'

Time was flying, and Carl had to report himself as arrived and enrol himself in one of the thirteen 'nations' of the university. He had to find himself in lodgings and settle down: all this to do before dusk—and the days shorten in September, especially so far north as Upsala. This palpable fact startled the young Smålander. He briskly returned to the town across the broken turfy ground behind the castle, through the court (that was to have been a quadrangle, only it was never made so, the town front of the castle alone being built) containing the long-bearded bust of Gustavus Vasa mounted on four cannons. He scrambled round among the unfinished turrets, finding no path down the steep hill on that side, but lingering a moment to behold the panoramic view extending all over Upland, comprising Old Upsala, with its 'Assize Hill' and the three tall tumuli known as the Tombs of the Kings.[1] Geologically, too, this view is very interesting, and Carl already knew all the geology that was then known in Sweden, and was constantly discovering more. The period following the glacial epoch was that of the roll-stone or sand-ridges. Such ridges are very common in Sweden; the celebrated mounds at Upsala are situated at the end of such a ridge.[2] This landscape is studded by Danmark and many other towers and villages, the Fyriså river gliding through the broad meadows which fade far off into the infinite ring of blue. He made his way round in front of the tall pinky-red castle where the green hill, which is here almost a cliff, makes a magnificent pedestal to a palace, and down into the university quarter—if one may say so of a town which is all university; though the academic buildings, looking like large private houses, are principally grouped on the north-west side of the town, just beyond the cathedral precincts.

While Linnæus is settling his affairs let us glance at the stately cathedral that he, who cared nothing for art, so heedlessly passed by. His own monument adorns it now, and his remains lie there in honour; but his monument was the last thing he would be thinking of just now—it was his work that lay before him. Victory before Westminster Abbey! Upsala Cathedral,

---

[1] They are each 58 feet high and 225 feet in diameter.
[2] Du Chaillu.

if less imposing than that of Lund, is a fine building, harmonious with the city's name of Upsalir, 'the lofty halls'; and made grander by being built on a height,[1] in a commanding and picturesque situation. It is approached from the main streets bordering the canal by a flight of steps leading through an archway framing delightful pictures of the Peasant's church, more ancient than the cathedral, and other buildings and parks, most of them connected with the university.

The cathedral is a very interesting building and full of charm; but before hearing mine or any other traveller's ravings—for travellers always come back raving from the North, though most of them do not intend to visit Sweden again: 'the sea-sickness is too horrid'—let us hear Fergusson, who never raves over Swedish buildings. He begins his concise account of Scandinavian architecture thus: 'No one who has listened to all that was said and written in Germany before the late Danish war can very well doubt that when he passes the Eyder, going northward, he will enter on a new architectural province. He must, however, be singularly deficient in ethnographical knowledge if he expects to find anything either original or beautiful in a country inhabited by races of such purely Aryan stock. If there is any Finnish or Lap blood in the veins of the Swedes or Danes, it must have dried up very early, for no trace of its effect can be detected in any of their architectural utterances; unless, indeed,

[1] Mons domini.

we should ascribe to it that peculiar fondness for circular forms which is so characteristic of their early churches, and which may have been derived from the circular mounds and stone circles which were in use in Sweden till the end of the tenth century.'

Does this solve the hard fact of Linnæus's coldness to art—that he was purely Aryan? But surely the Greeks were Aryan too. Or, does Fergusson, having hit upon an idea, knock his head against it too hard? 'The cathedral of Upsala can scarcely be quoted as an example of Scandinavian art, for when the Swedes, in the end of the thirteenth century (1278), determined on the erection of a cathedral worthy of their country, they employed a Frenchman, Étienne Bonneuil, to furnish them with a design and to superintend the erection, which he did till his death. After Bonneuil's death the French principles of detail were departed from.' The university buildings are not individually remarkable, although their grouping between the quaintly simple lines of the towered castle on its commanding hill and the rich Renaissance twin towers of the cathedral spiring up the valley, together with the undulating and well-planted slopes of the ground on the northwestern bank of the river, makes up a very pleasing prospect, with many picturesque points for the memory to retain.

Carl's immediate professors were Olaus Rudbeck (junior) and Roberg, both old men. Under them he made rapid advances in the different branches of medicine and natural history; and, regardless of the fact of his bread depending on the name he might win in the regular line of study, he revelled in all the gratifications of intellectual luxury. Life was one sparkling delight.

His was a 'bright, healthy, loving nature, enjoying ordinary innocent things so much that vice had no temptation for him.' Chief among his enjoyments—we know it from his remarks in after life—was to sail up and down the river to the Almare Stäket, where an arm of Lake Mälar narrows itself into a more riverlike branch, until it actually becomes the Fyriså River, flowing through Upsala.[1] The Mälar resembles a great sea-anemone, with arms in all directions, only that these arms have other arms, and so *ad infinitum*: a deeply pinnate fern-leaf is a more exact comparison. To sail through the Mälar is like seeing theatrical scenery unfolding as the capes and islands retire and disclose other islands, and beauties of lake and shore.

The Almare Island, with the ruined castle of St. Erik's Borg, stands in the middle of the strait or Stäket, where a swing-bridge lets the boat pass into the long and sleepy Skarfven, as this arm of the lake is called; the shores are lined with gambrel-roofed cottages set in foliage; boats, fishing-nets, and good agriculture enliven the soft and soothing landscape. Then comes the red-roofed town of Sigtuna, which has gone so completely to sleep these last seven hundred years, that

[1] There are no lakes immediately by Upsala.

after being one of the largest and handsomest towns in Sweden, it is now a mere village with picturesque ivy-covered ruins and five hundred inhabitants. The massive silver doors of one of its churches were carried off in 1187 by the Esths to Novgorod, where they may now be seen. Numbers of the white-capped students of Upsala are to be met cruising about this part of the lake and river nowadays. Rail and steamer both aid their peregrinations.[1]

The grass here grows vividly green, and the trees are fine and flourishing. Presently (we are sailing towards Upsala) the foliaged banks subside to a narrow low-shored arm of the lake, looking like a shallow river rather than a lake, lying between water-meadows, until the banks rise again into fir woods and hills clothed with silver birch and the foaming white-blossomed bird-cherry—essentially a Linnæan tree, as it always grows abundantly round his dwelling, wherever that may be.

At Skogkloster—once a forest monastery, now a fine square chateau, with copper-roofed towers at the angles and much magnificence within—the water widens out into a broad bay on the left bank, and the channel turns off to the right towards Upsala. The water's local colour is a greenish brown or warm olive, lowered by the sky reflections to a neutral grey. The boat here

---

[1] The undergraduates wear a white cap with a black velvet band and a small blue and yellow rosette in the centre, symbolic of the Swedish flag.

enters the Fyriså through a drawbridge and by a chain of low islets with wind-tost and water-washed fir-trees growing on them. The river, shallow, muddy, and rush-banked, is rich in water-lilies and marsh plants. The green slopes shelve gently upward to the Scotch fir, spruce, and pine trees, which feather down to the grass: the banks are so rotten that they are worn away by every wavelet. The fine modern agricultural school here has not devoted its attention to the first principle of riverine agriculture, the solidifying of the land-banks. The mud dissolves like sugar behind the passing steamers, and is swept down in rich liquid form, to settle at the bottom of the Mälar.

The boat sweeps by the famous 'King's Meadow,' which Linnæus afterwards so loved, which in spring is one carpet of fritillary, chiefly purple, mingled with the white and red sorts. High above this historical scene rises the round red tower of the Slott, or Castle of Gustavus Vasa, with the Dutch-looking town of Upsala lying at its foot, and the long stretch of canal-like river is closed in by the lofty brick cathedral reflected in the pools between the five bridges.

Carl, free from care or anxiety respecting his bodily support, worked with all possible zeal. He had one great disappointment, however. The greatest adept in natural history, and especially in botany, in Sweden, was Olaus Celsius,[1] the first professor of Divinity, and dean of the chapter of Upsala. Linnæus described him

[1] Olof in the Diary.

later in a letter to Haller [1] as the only botanist of his country, and Carl had hoped to profit by his learning. Celsius at this time was away on official business at Stockholm, so that Carl was obliged to continue his favourite study with no guidance save that of his own genius and the works of the men of the last two centuries—in fact, the same materials that Celsius himself had; but he was minus Celsius' years of experience.

A year passed. With his vivacity of temperament, he could not manage his small finances to advantage—he was too sanguine—he felt too sure of immediately conquering fortune somehow. Well as he had been trained in economy, it is difficult to square 8*l*. with a journey, clothes, board, lodging, and tuition for a year. It is not very easy to do it for 40*l*. by a popular fellow, naturally open-handed, whose pleasant speech, and face beaming with frankest good-humour, made him courted by the pleasure-loving youths of the university.

A short time before Carl came northward, his rival at Lund, Nils Rosen, had been appointed adjunctus of the faculty of medicine at Upsala; he laughed at Linnæus's hopes of that pension for his talents which Rothman had encouraged him to look for. The professors looked coldly on one who brought no ready-made reputation with him, and who seemed unlikely to do them credit, as he only pursued an inferior or incidental branch of learning—for botany, until Carl made

[1] Dated from Hartecamp, near Leyden, May 1737.

a profession of it, was not a profession at all. He was finding his level among them, the students thought, and Rosen said it. 'Young Linnæus always had too good an opinion of himself.' One would say, we often do say, that to know the marvels of creation keeps one humble; yet, somehow, young scientific men are seldom humble, in expression at least, and Linnæus was no exception. But he did not often get a snubbing,[1] nor were his days sorrowful, though he had not yet set the Fyriså on fire. Now were his joyous friendships, his pleasures of hope. Carl gave little heed to Rosen now: he was absorbed in a deep friendship. Hear the beginning of it in his own words.

'In the year 1728,' says Linnæus, 'I came to Upsala. I asked what student was most eminent for his knowledge in natural history. The name of Artedi was heard everywhere; he had studied there several years before me. I felt the most ardent desire to see him. On paying him a visit I found him pale, downcast, and weeping because his father had just died. Our conversation turned on plants, stones, and animals. The novel remarks he made, the knowledge he displayed, struck me with amazement. I solicited his friendship, he wished for mine. How valuable, how happy was our intercourse! With what pleasure did we see it cemented! If one of us made a new observation he communicated it to the other; not a day elapsed without our receiving reciprocal instruction. Rivalship

---

[1] A Swedish word; *snubba*, a rebuke.

increased our diligence and researches; though we lived at a great distance, yet it would not prevent us visiting each other every day. Even the dissimilitude of our character turned out to advantage. He excelled me in chemistry, and I outdid him in the knowledge of birds and insects and in botany.'

Artedi also studied alchemy: the poor youth added the golden dream to that of the lordship of creation. Peter, or Pehr, Artedi was two years older than Linnæus. He was born in 1705 in Angermania, likewise of poor parents. His behaviour at the college of Hernösand was the counterpart of our Carl's at Wexio, preferring the study of nature, especially that of fishes, to all other accomplishments. In 1724 he came to Upsala to study divinity, but he soon exchanged theology for natural history.

Pehr Arctedius—for this was his name in Swedish, only he shortened it to Artedi—went to Angermanland to discharge the last duties to his father, and on his return gave himself up to the pleasures of a friendship with Linnæus. Artedi was of a tall handsome figure; Linnæus was shorter, stouter, more hasty in temper, and fuller of youth's certainty of success. They both had a noble spirit of emulation; they were 'opposing mirrors, each reflecting each'; every discovery or thought returned to each improved, enlightened by passing through the other's mind; the flashes of illumination were caught in talk and fixed.

Enthusiasm is catching. Carl's flame fired Artedi

also. 'As soon as one found himself unequal to the progress of the other in one species of study he dedicated himself to another. They therefore divided the kingdoms and provinces of nature between them.'[1] They began to study insects and fishes together, but in a short time Linnæus yielded the palm to Artedi in ichthyology and the latter acknowledged Linnæus to be his superior in entomology. Artedi undertook to reduce amphibia, and Linnæus birds, under a regular arrangement. Each kept his discoveries to himself,[2] though for no length of time, since not a day passed without one surprising the other by narrating some new fact.

Artedi finally confined his botanical studies to the umbelliferous plants, in which he pointed out the distinction which arises from the differences of the involucrum, leading to a new method of classification, which was afterwards published by Linnæus, with a tribute to his friend. 'But the chief object of Artedi's pursuits, which transmitted his fame to posterity,' says the rapturous old Stoever, 'was the empire of Neptune, or the knowledge of the natural history of fishes, called ichthyology. Linnæus relinquished to him this province.' Emulation is the soul of improvement. Laying their plans so as to assist each other in every branch of natural history and medicine, Artedi had projected the happy plan of introducing a new method and classification in ichthyology, which cheered and strengthened Linnæus to effect the same thing in botany. They 'worked

[1] Linnæus.   [2] Diary.

deliciously hard ; felt light, happy, invincible '; and they loved like David and Jonathan.

To Artedi Linnæus was like a young brother—like himself, but more ardent : as Frederika Bremer says of another naturalist (Kingsley), 'a young mind that he could like, love, quarrel with, live with, influence, be influenced by, follow through the thorny path, through tropical islands, through storm and sunshine, higher and higher ascending into the metamorphosis of existence.' Both were handsome in feature, improved by the beauty of expression caused by the habitual admiration of God's works : the love of beauty, and of God who made such beauty, passes into the countenance and glorifies it. Their faces were thus habitually cast in noble lines, animated by the eagerness of innocent discovery. They had no lower lusts, poverty kept them from all other indulgence, disciplined them. They had none but intellectual pleasures, and these of a fine kind. At first they laughed at poverty—they, so rich in gifts, health, youth, affection, admiration, all that makes life so precious.

> Dans un grenier, qu'on est bien à vingt ans.

Earth, air, and water were full of their familiar friends. They daily sought and found that beauty which Plato defines—it goes best in French—' Le splendeur du vrai,'[1] while Aristotle as truly declares that beauty consists in the complete development of beings, each according to its sort and nature,—the groundwork of all science. The

---

[1] Which, indeed, is the best definition of Art.

two young men lodged far apart. Artedi naturally preferred the situation by the river-side, below the castle hill and the present hospital, where the Stromparterre is now, where the band plays of an evening; while Linnæus chose to be nearer the botanical garden and the museums. Sometimes they met farther down the river by the flowery 'King's Meadow,' where the water-byssus grows in ditches by the wayside, particularly in places sheltered from the wind. 'It resembles the cream of milk,' Linnæus says, 'and is called by the peasants the water-flower.' Here both were best suited

Sometimes they would be seated on the moss-tufted castle slopes, where grows the rare moss, the *lichen nivalis*,[1] looking away over the distance, far-reaching as their fancies, talking of the future; where often also the two elderly professors, Rudbeck and Roberg, might be seen, as the professors may frequently be seen at this day, pacing the grassy terrace in front of the castle, not exactly arm-in-arm, but the taller with his arm around the other's neck, the shorter holding the other round the waist—a sight queer to English eyes, but which passes perfectly unnoticed here in Upsala: these would be talking of the past. And what a difference in the ideas and the talk of the two pairs! Contrast the seniors' converse, tough and sententious, with the burning young ideas or the limp new-born ones coming forth copiously and with every form of expansion. Yet there was occult talk between the juniors also, Artedi groping

[1] Linnæus.

his way in the unutterable and sublime; Linnæus more practical, eager for praise and profit. Their minds, if raw, were receptive. The elders' were closed to any new discoveries: memory was broad enough for them. These two old professors could not sympathise with the young men, but Celsius would come soon, they reflected, and Celsius would understand them.

Young people think our old inheritance of ideas, our civilisation, our religion, and our principles are ancient petrifactions. They are not so; rather are they like wood, old yet alive, from which spring the shoots, the leaves, the sprigs, that look so different. They will become the same: intrinsically they are the same.

Sometimes the youths would dart down the steep slopes and chevy away in the far distance in chase after a bird or beast, or something attractive viewed miles away. Both were swimmers: most Swedes are so, and have need to be in that lake country. There was no end to Carl's feats of agility in rock or wall-climbing, and of adventurous courage to get birds' eggs from orchard, cliff, tree, or tower; unwearying his zeal, that never felt fatigue while in the chase, by night or day. Of happy disposition generally, Carl was of quick temper; his anger was violent, but soon over; though he would sometimes be chafed to exasperation by a seeming trifle. He loved the hardest study, laboriously travelling in search of facts; not careering his mind through fine districts—the villas, parks, and esplanades of classic lore —but changing ancient unreal dreams for facts, he fought his way through difficulties in unknown or fresh-broken ground.

Though genial in temperament Linnæus cared little for athletic sports. Perhaps few Swedes do. I have seen Swedish boys at brisk play in the gravelled or pebbled squares in front of their grammar schools; but games do not seem to thrive among them like football does with us. They are such long years behind us with their tools—their bicycles, for instance—that as we laugh at their 'wobbling' movements we forget how we grinned at our own early velocipedes. They play croquet too, now that it has been for some dozen years superseded by lawn tennis.

Carl's favourite haunts were beyond Danmark Church and the ten ancient Mora stones, round by Hammarby and Sofja, where the clay soil of Upsala Vale changes into the heathland of the hills consisting of sand and stones; he was reminded by these glacier-worn rocks of his home in Småland. Here he could revel in discovery; here he felt those glorious moments when the soul, risen by hard-won ways mountains high, overlooks the fair world of common things in the clear air, the second heaven, of purity. He prolonged the comfort of these excursions to the latest autumn, 'those seasons of silence and twilight when nature seems to sympathise with the fallen . . . to soothe and comfort, to inspire and support the afflicted.' For as time went on and a second winter was approaching—a Swedish winter—and yet appreciation came not, bringing scholarships and all fat things, it seemed as if the corporeal portion of the complete development of these two poor geniuses were at a standstill. Petronius says—it is Linnæus who quotes him here—' Poverty is the attendant of a good mind.' 'Never mind,' said Carl, cheerily quoting a local proverb, ' put a Smålander on a barren rock in the sea, and he will manage to make his living.' Artedi shook his head. Less hopeful than Carl, Artedi was pensive and sentimental, and susceptible of soft emotions. Philosophy is much, but it is not bread and butter. Carl's pockets were quite empty, and he had no chance of obtaining private pupils, who, in fact, are seldom put under the care of medical students. It is said he obtained on December 16, 1728, a royal scholarship, of the value of which we are not informed,[1] but which was quite insufficient to maintain him. Stoever denies this, and it seems doubtful. The Englishman has perhaps confounded this with a bursary he really did afterwards obtain—Wrede's exhibition, value about 5*l*.

The woodland soft fruits were all over; the nuts would soon be gone too, and the edible roots that the two friends knew so well how to find in summer; the fish, too, that they caught, examined, dissected, cooked, and ate with their rye biscuit, would soon all be locked beneath the ice, as winter fell 'a heavy gloom oppressive o'er *their* world.' Hitherto they had relished their plain living and high thinking while, over some old book recently ferreted out of the lost corners of the

[1] Smith.

library, or some fresh winged thing discovered in the air, they seasoned their spare dinner with proverbs either national or of their own coining, bracing up their soul with maxims, persuading themselves that the wants, anxieties, privations of life were nought when set against the endless rapture of perpetual effort to realise a grand conception.

Had we means answering to our mind.[1]

'Nothing like poverty for strengthening the character,' would Artedi say, capped by Carl with 'Many things are more precious than a full stomach,' and his friend's rejoinder by-and-by, that ' royal roads do not make a great people.' Yet the burden of their inmost feelings was 'Oh for Celsius! Oh, if Dean Celsius would but come!' If he came their talents must be recognised. 'Alas, good and quickly seldom meet,' said Artedi, with the ready proverb's ' deep though broken wisdom.' The aged medical professors, Rudbeck and Roberg, were limited and dull, and little inclined for improvement, which meant movement; and old men are disinclined to stir. These men were pamphletary rather than practical; but Celsius was still in the prime of life and zealous for his favourite science.

Linnæus felt his woes deeply aggravated by Celsius's prolonged absence, as his coat became more and more frayed at the seams and edges, and threadbare. For all their tall talk about the royalty of science, it was hard when Rosen stalked by neatly dressed, or was seen

[1] *Paracelsus*, BROWNING.

sitting at the windows of the Stadhuset dining with a professor; it was hard to feel that they two would be known for the handsomest young men in Upsala had they but had new coats and white silk caps. Carl's shoes, too, he had had them soled thrice; he thought he had better have upper leathers put to them this time for a change. At last nothing was left of them but the strings, he tramped so much; and there were irreparable defects in some parts of his equipage which could not be concealed by 'all sorts of coaxing, darning, or sitting cross-legged.'

> A blasted bud displays yon torn
> Faint rudiments of the full flower unborn.—*Sordello.*
> But who divines what glory coats o'erclasp of the bulb dormant.
> *Paracelsus.*

There was a certain grim humour in seeing these two ragged students portioning out the animal, vegetable, and mineral kingdoms between them; dividing, as the Romans had done, the domination of the world.[1] They who could not buy an oxstek[2] or a juicy turnip in it; yet were they victors. Like Alexander, they had whole provinces to their hand in little—epitomised—on their study shelves: collections of rubbish, valueless in themselves, valuable in their classification; their mineral collection, complete in granite, and gravel, and ironstone; the cells for gold and silver empty; rubies would come by-and-by; meanwhile there was their place ready and their analysis neatly written out. As Elia says of Captain Jackson, 'with nothing to live on

---

[1] *Baeck.* [2] Beefsteak.

he seemed to live on everything. He had a stock of wealth in his mind—not that which is properly termed content, for in truth he was not to be contained at all, but overflowed all bounds by the force of a magnificent self-delusion.' Artedi took to himself the realm of fishes, which Linnæus willingly ' conveyanced ' to him ; but when Artedi required a province of his friend's own particular kingdom, and wished to take the umbelliferous plants under his rule, this was a harder concession to friendship.

The two friends were always finding something fresh, acquiring property too—a treasure-chest—but of a sort whose key was in their mind. There is nothing like having little or no cash for making one's collections of value. One buys no trash, nothing that salesmen of curiosities consider suitable for amateurs. One gleans, not from books, but from the substances around, completing an area, exhausting the neighbourhood, from its chalk-hills to its clay-beds. Each saw himself in the glass of his friend's admiring mind, and each felt comfort in the possession of commanding talent. They *must* rise, and they would.

' Or, staggered only at *their* own vast wits,' no wonder if these two students felt stuck-up, over-elated at times, when they considered the education the rest of the fellows were getting in the university. Professor Rudbeck exhibited to the students his beautifully coloured drawings of birds, and Professor Roberg lectured on the problems of Aristotle according to the principles of Des

Cartes. In anatomy and chemistry there was profound silence; neither did our botanist ever hear a single lecture, public or private, on the study of plants.[1] Oh, when would Celsius come and disperse this gloom, stir this stagnation, and begin to teach?

'During this period of intense receptivity'[2] Linnæus read in the Leipsic commentaries a review of Vaillant's treatise on the sexes of plants. Here was a ray of light. Oh, for Celsius to come and help him to read by it!

Linnæus was beginning his second year at the university. His pockets were empty; subsisting on accidents, he picked up a meal here and there by helping duller students, and from their charity. He learned by heart that marvellous lesson in natural history, that 'of all God's creatures, man alone is poor.' Now his clothes gave way completely, and winter was coming on. Winter begins to bite early in Sweden. Carl, who was proud of his personal appearance, and had always taken pains with his dress, was now glad to cover himself with the cast-off clothes of his more wealthy companions. He grew used to 'the mean and bitter shifts of poverty,' and gaunt and haggard with actual famine.

He often spoke of this in later life (as well as in his installation speech in 1741 as professor at Upsala), telling how under severest poverty he could return thanks to God whose Divine Providence guarded and

---

[1] Stoever.   [2] Jackson.

supported him. He thus made his own case an encouragement to other poor students, and also a lesson in patience; for victory does not come with a leap—her path must be laboriously prepared.

He put cards and pasteboard in the worn-out shoes given him by his comrades, and stitched and mended them with birch bark, neatly and carefully, for he was neat-handed with his glueing of plants and preparation of specimens—a good thing for him, for, as George Eliot wisely says, 'Some skill with the hands is needful for the completeness of life, and makes a bridge over times of doubt and despondency.' The lowest price of a pair of common boots was nine (copper?) dollars, and of strong shoes five dollars.[1] He thought, as he sat mending his shoes, that perhaps the cobbler's trade had been a better life after all. This brought to memory his father's kindness. He felt like the repentant prodigal—I will arise and go to my father. But no, his father could not help him—his parents had too many mouths to feed; he would not sponge upon their small store.

He would gladly have returned to Stobæus at Lund, but Stobæus had taken it ill that a pupil whom he had treated so kindly should have left the university without consulting him.[2] No, he must win his way upwards by himself; and as Artedi saw the conqueror shine through the darkened splendour of his eyes, he sighed that he himself had not the same victorious constitution, that he could not equally pull the chariot of science.

---

[1] Linnæus's Lapland diary.   [2] Stoever.

Meanwhile cold and hunger both grew harder to bear; 'the owl, for all his feathers, was a-cold'; and in the depth of a Swedish winter, where to study one must also burn the *midday* oil, Carl could buy neither candles nor oil for his study lamp. Winter was Linnæus's especial enemy, putting ice for minerals, shrivelling his flowers to dust, and leaving him thin as his own darning-needle. Where a good-natured friend gave him a light, it was a sacrifice to burn the rare and luxurious candle that he might have eaten. What a conflict between the bodily and mental appetite!

When Dr. Johnson said 'the distinction of seasons is produced only by imagination operating on luxury' he had not felt a Swedish winter. What sounds wise and sensible when said over the second bottle of port at the 'Mitre' is less true when, empty of pocket and of stomach, one shivers in thin garments outside the tavern. Skating is glorious exercise, but one cannot even slide, Sam Weller fashion, barefooted. When one has sewn one's boots with birch-bark and pasteboard, one is as careful of them as Don Quixote was over the second edition of his helmet.

> Nothing in poverty so ill is borne,
> As its exposing men to grinning scorn.

The Swedes are too polite to sneer at even unprofessional cobbling, and Carl carefully stitched and mended his best shoes so that returning daylight might at least enable him to go out and gather plants. Spring was opening up after the long and bitter winter, when cold and famine fought over his body; when even his mind starved in those noontide twilights, without even a rushlight to warm body and soul by; when at night he would shiver for hours till he fell asleep. Each day at dinnertime he felt the want of the meal; and though he at first fought this off by trying to absorb himself in a book, he found his mind wandering through faintness, and he had to go and lie down till the hunger pang passed off. Carlyle, in the inflated style of his youth, feelingly says, ' Few things in nature have so much of the sublime in them as the spectacle of a poor but honourable-minded youth, with discouragement all around him, but with never-dying hope within his heart; forging, as it were, the armour with which he is destined to resist and overcome the hydras of this world, and conquer for himself in due time a habitation among the sunny fields of life.' The ancient Scandinavian spirit within him made Linnæus 'firm to inflict and stubborn to endure.'[1]

But the broad blaze of summer now coming, when even beggars might be fed cheap and warmed for nothing, would be all the more radiant for the long howling darkness of six months. Even this dreary winter stage had been sweetened to Linnæus by youth's hopes and friendship, the sweet savour of life, the peculiar boon of heaven,

> To men and angels only given,
> To all the lower world denied.

The friends inspired and warmed each other with fine

[1] Southey.

words, and such visions Linnæus could raise by his eloquence in the mind of his hearers, his Pylades, his 'solitary luxury, his friend,' that at times they even grew drunk with the wine of their enjoyment. But

> Fate is tardy with the stage
> And crowd she promised. Lean he grows and pale,
> Though restlessly at rest.—*Sordello.*

Sadly Carl munched his rye biscuit by the warmth of the stoke-hole fire in the winter plant-house, and still waited.

> The woods were long austere with snow; at last
> Pink leaflets budded on the beech, and fast
> Larches, scattered through pine-tree solitudes,
> Brightened . . . Our buried year grew young again.

Carl, who had sighed for Celsius to come for his instruction merely, now looked to his coming as his only possible salvation. How, he knew not; but with all the faith not yet starved out of him, he was sure that a new life for him would at once begin. He was now in debt for his lodging, and debt gnaws sharply.

At last, in desperation, and by the advice of Professor Roberg, Carl applied for the situation of gardener in the academy gardens; but this was refused him by Professor Rudbeck, who remarked at the same time that he thought him qualified for a far superior station.[1] He says 'he repined very much at this denial.'

Oh, if he could but be free of debt he would forsake all his hopes, all his dreams; he would leave Upsala, leave his friend, and go home and be obscure! With

---
[1] Diary.

energies burning themselves out unused, anxiety, worn-out hope, and leanness preyed upon him. He did not know what it was to have a full meal. Bitterest of all to him was the sense of failure. There was one lower step. If our Johnson felt savage as he did when some well-meaning clumsy person put new shoes outside his door, what must Linnæus have felt when Rosen, who was now going abroad for the purpose of improving himself and obtaining his Doctor's degree (which by the Swedish rule must be taken in some foreign country), left him an old but respectable suit—Rosen, who had despised Linnæus in his rags!

'I would rather die than put it on,' cried the fierce Linnæus. In debt though he was, he could not be indebted to Rosen. The excellent Rosen complacently thought of coals of fire. Rosen went abroad, and by-and-by became a distinguished man at Upsala; he was ultimately ennobled as Von Rosenstein. In the meantime his place as adjunct teacher was supplied by an incompetent student named Preutz. Linnæus, bound by poverty and chained by debt, could not leave Upsala even to become a mechanic in Småland, not for all the flowery language in which Stoever talks of his taking leave of the Muses and of the goddess Flora. But he endeavoured to do so: he made a determination: he would beg money of his father, of his relations, of Stobæus; he would so far humble himself, and leave Upsala and the bright future he had failed to conquer.

Oh, for Celsius! Oh, why had Celsius never come?

CHAPTER V.

DEAN CELSIUS COMES.

> Thou seek'st in globe and galaxy,
> He hides in pure transparency.
> Thou ask'st in fountains and in fires,
> He is the essence that inquires.
> He is the axis of the star;
> He is the sparkle of the spar;
> He is the heart of every creature;
> He is the meaning of each feature;
> And his mind is the sky
> Than all it holds more deep, more high.
> *Woodnotes*, EMERSON.

LINNÆUS was about to quit Upsala, when, standing one morning in the garden he loved so well, before a newly-opened flower—one he had never seen bloom before—'I will cut it,' said he—' a last specimen for my herbal, a "**minné**" of happy days gone by—and then depart.'

Carl stood not in the garden alone: a voice answered from behind, 'You will do no such thing; leave the flower.' It was the **professor** of Divinity at Upsala, in gown and ruff; it was Dean Celsius himself, but Linnæus did not know him.

Old Stoever gives another version of the tale. I prune it of some of his exuberant syllables. 'One day in the autumn of 1729' [it was, in fact, early summer], 'while Linnæus was intently examining some plants in the academic garden, there entered a venerable old clergyman' [Stoever always adds the picturesque touch, but Celsius was just forty-nine.[1] Is that such a venerable age?] 'who asked him what he was about, whether he was acquainted with plants, whether he understood botany, whence he came, and how long he had been prosecuting his studies.

'Linnæus answered all these questions, and, when his interlocutor showed him various plants, mentioned their names agreeably to the system of Tournefort. Being further asked what number of specimens he possessed, he replied that he had above 600 indigenous plants preserved in his cabinet. He was requested to accompany the gentleman who had thus interrogated him to his house, which proved to be that of Dr. Olaf Celsius, and the interrogator was the Doctor himself just returned from Stockholm.'

The Dean spoke kindly to the youth; Linnæus trembled like the aspen. Intuition told him who this was. Had Celsius, had Fortune really come at last? Carl's thin cheek reddened, his eyes filled at the tone of kindness. Hot tears, a choking sensation in the throat, came at the words of encouragement from an elder; the first for so long. Youth is always so hungry for kindness, and Carl was used to wanting—bread. The ragged youth spoke of the plants to Celsius, describing them

[1] He was born 1680.

with an exactness surprising in a student, and upon nearer conversation displayed such extensive knowledge as struck Celsius with astonishment. At last the sun had risen upon Linnæus. They talked; the dean listened with interest while the young man spoke with an enthusiasm which for the moment sent the rich blood of health into the student's pale features, long since wan with insufficient food. His threadbare clothes and patched shoes told their own tale; 'starvation wrote as a notice-board on his hollow cheeks, skinny fingers, and sunk eyes, went straight to the heart.'[1] Soap costs little and water nothing in Sweden, and manners come by nature: the gentlemanly bearing and the exquisite personal cleanliness of Linnæus made him known for a gentleman at once; all the rags in Upsala could not disguise the gentlemanhood of the man refined by loving all things lovely.

Carl had an agitated walk to Artedi's lodgings. His eyes glittered with excitement as he told the good news to his friend. Now they should both get on: he would give his friend a helping hand. How volubly they talked! It was as good as a full meal to both.

Inquiries were made. Celsius heard of Carl's distresses and his inoffensive mode of life, and the dean took him into his house and was ever kind to him, and made him tutor to his younger children. The advantages were mutual. Celsius too had found what he wanted. For thirty years he had been intent upon

[1] Sam Slick.

illustrating the plants of Scripture, and, himself an
adept in Eastern tongues, had travelled to the East to
inquire into and study these plants in their native soil.
He was now at work preparing his ' Hierobotanicon : a
Critical Dissertation on the Plants mentioned in Scripture,' only needing some more youthful help to make the
work perfect and bring it before the world. Now had come
the hour, the man, and the collaborator ready made to
his hand. 'There is no education like adversity': one
readily turns one's hand to anything. Linnæus bore an
active share in the production of this learned work,
which is in Latin, and, alas, sadly fails in interesting
the ordinary reader. It was published in 1745 and 1752
in two volumes. As there were only two hundred copies
printed, the book is of course now very rare, which is as
well. The dean could not even interest his eldest son in it;
but then he was working at his own line of mathematics.
The arrival of this young man, just six years older than
himself, was a great additional pleasure to Linnæus.
They became firm friends. Andres Celsius is one of
Sweden's most celebrated men. Later he joined Maupertuis and his associates in the measurement of the
Lapland degree, and afterwards built an observatory at
Upsala. He was the first who employed the centigrade
thermometer. He wrote astronomical and meteorological observations and a collection of the Auroræ Boreales
observed in his time in Sweden.[1] Linnæus, however,

[1] Under the title *CCCXVI Observationes de Lumine Boreali*,
1733. Nuremberg. Andres Celsius, born 1701 at Upsala, died 1744,

took a deep interest in the ' Hierobotanicon ; ' as we shall see later on, it was a subject on which he felt keenly as the central point of botanical study, comprising as it does objects of such vital necessity and convenience to mankind. This book of Celsius' could never have satisfied Linnæus, who would have liked a complete Flora Palestina with all the plants that Scripture does not mention by name as well; but it gave him an insight into the way of preparing such works, and made him ambitious himself to become an author.

The sap rose in the frozen body : it was the springtide of his life, and, as usual, the epoch of creative power. Carl had already composed a little catalogue of his botanical observations, under the title of 'Spolia Botanica,' Upsala, 1729.[1] This was never published. The original, written in Swedish, is preserved with the collection of MSS. brought to England by Sir J. E. Smith. It was dedicated to Professor Roberg, and contains sketches of a few of the plants, arranged on Tournefort's system, and a rude map of their habitat.

must not be confounded with his father Olaus Celsius, 1680 (some say 1670) to 1756, theologian and botanist ; or with his grandfather, Magnus Nicholas Celsius, 1621–1679 (?), mathematician and botanist. Linnæus probably accompanied the younger Celsius in a rapid visit to Dannemora, of which we find the only trace in the dates in the note-book previously mentioned : ' Journey to Dannemora, May 24 ' [he had just kept his birthday], ' 1729 ; June 10, travelled to Upsala.' No biographer of Linnæus mentions this expedition. It must have been very soon after his appointment with Celsius. The dates of his life at Upsala present many difficulties, Stoever, the diary, and the note-book are so contradictory. Even Linnæus's own written dates do not always tally.

[1] Pulteney.

The arranging of the 'Hierobotanicon' was one of the chief motives which made Celsius take Carl into his house (though he afterwards became like a son to Celsius, and he a father—a true adoption). For this purpose also he had free use of Celsius' library, one of the richest and most valuable in Sweden. Here Carl now met with Vaillant's small treatise on the sexes of plants,[1] a review of which he had already read in the Leipsic commentaries; which gave him the first notion of the sexual distinctions of flowers, the groundwork of his celebrated system; which, after all, contains the spark of a grand illumination. Hitherto he had worked on Tournefort's lines of classification by form.

This was the germinating moment of his life. To many of us it happens to be once, if only once, struck dead, as it seems, to all outward things by the lightning shock of an idea. This flash is what must be brought to crystallisation by hard and continuous labour. This idea immortalised Linnæus's name, and deservedly so, since although this was no new notion—that of sexes of plants—Linnæus first applied it to classification and elucidated it. 'Principles had to be imbibed in copious draughts all through his education. The collision, combination, harmonising of these constitute speculative insight and conduct to original thought.'[2] The Linnæan is an artificial system; its author saw that fact as plainly as we do. But, though imperfect, it was a high road towards a new method of thought.

[1] *Sermo de Structura Florum.*   [2] Bain on Mill.

It was a new doctrine in his day, and brilliant in its brand-new gloss, although derived from a hint as old as Aristotle, who gives this glimmering—'If the dust of the branch of a male palm be shaken over the female tree, the fruit of the latter will ripen quickly.' Le Vaillant was not the first to read Aristotle, but he was the first to apply the idea to flower-structure, to the pistils and stamens of plants. One can go further with other people's ideas than with one's own, is a saying true all the world over: it was left to this young Swede to take the leap from Le Vaillant's standpoint and bring the long-desired system out of the obscurity beyond. ' Till now Linnæus had considered plants by their bloom, hitherto the stamina and pistilla had been considered insignificant.' A mere finish to the beauty of the flower, a fringe and tassels. ' The idea of a better system than that which Vaillant had hinted now guided his botanical observations.' As in human nature families are named from their marriages, so with plants he would make this the basis of nomenclature. The further he brought his theory forward the more consistency did he discover in his own knowledge, the more powerful were the attractions of the plan. Oh, the fear lest someone might forestall him! And this alarm was not unfounded; for though a truth may have lain dormant for thousands of years, yet the moment the earth is ready for its appearing it will spring up, and someone will, and must, be the first to light upon it. 'The sexes of plants now occupied his thoughts night and day.'[1] During this time of intellectual fever he kept his mind jealously aloof. He hugged his precious secret even from Artedi: their habit of keeping their discoveries close till perfected was of service to him now: he would wait until at last he could bring out his fair idea complete, clothed in a system, and show his new Eve to his bosom-friend, and then under four eyes only. Artedi was the first; next day it was unexpectedly public to all Upsala. A disputation was held before Bishop Wahlin on the 'Marriage of the Trees: sive Nuptiæ Arborum.' This was a blooming new idea in the summer of 1730.[2] Linnæus was present. The subject of the controversy was familiar to him. None found it more pleasant, nor had anyone at Upsala studied it better than himself.[3] Linnæus was in his element; now was his hour—the opportunity that comes once in life to all men. Even Artedi, his bosom-friend, was astonished at his radiance.

The account in the diary adds a few particulars. 'There was just then published a philological dissertation " De Nuptiis Plantarum "[4] from the pen of George Wahlin, librarian of the university; and as Linnæus had no opportunity of publicly opposing it, or of stating his doubts, he drew up in writing a little treatise on the sexes of plants, and showed it to Dr. Celsius, who put in the hands of Dr. Rudbeck. The latter honoured

[1] Diary.

[2] Glittering Darwin's *Loves of the Plants* delighted the reading world in 1789.—FREDERIC HARRISON.

[3] Notes for Biography Linn. [4] Or *Arborum*.

it with the highest approbation, and expressed a wish to be better acquainted with the author.

This small treatise, replete with new and luminous observations, delighted Professor Rudbeck; he was struck with the young author's spirit of observation and the solidity and novelty of his knowledge. Old Rudbeck was not altogether one of those professors 'miserable creatures lost in statistics'; he loved a theory dearly. He wrote paradoxes by the score, and a thick book of hypotheses to prove that all Europe was civilised from Sweden.

'We'll verify his words, eh, Artedi?'[1] said Linnæus. The young men used good-humouredly to laugh at the good old theorist. 'Rarely has such a variety of profound and extensive learning been united as in Rudbeck,' writes Linnæus. 'But he maintains the strangest and most unbounded paradoxes. He pretends that Sweden was the abode of the ancient Pagan deities and of our first parents; the terrestrial paradise, the true Atlantis of Plato; and that it was the origin of the English, the Danes, the Greeks, the Romans, and all the rest of the world.'

Linnæus was brought forward to dispute upon his

---

[1] The view now gaining ground is that the Aryans originated in Europe, say in North Germany or Sweden, that the Sanskrit-speaking conquerors of the Land of the Five Rivers were, in fact, the Eurasians of their time. Mr. Saporta's notion is that the human race originated within the Arctic circle at a time when most of the surface of the globe was too hot to be inhabited by man.—Hibbert Lecture, May 1886.

thesis, which he did in the most brilliant style. Without a copper dollar in his pocket, though no longer in rags, he was an object of great attention. This was life indeed. It was as if he had dropped from the stars, so little had he been recognised in Upsala. 'Thought is the soul of act.' He had prepared his soul in unrecognition, as all such souls must be prepared. Now he could expand, give wings to thought, ply act on act; build an edifice on what was once but a theory, like an architect's design set in accomplishment. To work out both demands outward influence. His was a fresh soul created, late in space, as the new stars are, when the world was ready to receive it.

Professor Rudbeck, under whom he had been principally working, was the most amazed. Celsius' swan, then, really was a swan. We can readily fancy the triumph of the worthy dean in having at once made the discovery that the other professors in over eighteen months had failed to make. Of course some envy was excited, but Rudbeck was too generous to feel piqued with either Celsius or the youth.

Most people know who Rudbeck was, but in case overloaded memory should confound him with a greater Olaus Rudbeck, his father, I will faintly outline the lives of both. Olaus Rudbeck, junior, born 1660, was the son of Rudbeck, an anatomical discoverer, or more like what we now call a comparative anatomist, protected by the clear-sighted Queen Christina of Sweden. The senior Rudbeck established the botanical garden at

Upsala. He travelled at the Queen's expense and collected a vast quantity of plants and herbs, most of which, and the greater part of his valuable writings, with nearly all the 1,000 blocks prepared for the engravings of his great botanical work, were destroyed in the great fire at Upsala in 1702. Oxford possesses some relics of this work, and the Linnæan Society a few of the engraved blocks. Rudbeck did not long survive the destruction of his labours: he died at Upsala, December 12, 1702, leaving his son, who had accompanied him in his Lapland travels, to carry on his work and repair if possible the havoc of the fire. Linnæus named a plant after him. The junior Rudbeck (with whom his father's dying wish was a pious heritage that he had never yet been able to fulfil) was now seventy, and going out and giving lectures were difficulties for him. He wished for an assistant. Rosen being gone, Rudbeck had hitherto employed his nominee, Preutz, to read his lectures for him; but his incompetency deprived them of all their spirit: a dull man himself, Preutz dimmed whatever he handled. The perusal of Linnæus's treatise, and further examination, determined Rudbeck to fix on him to replace Preutz. Accordingly, he invited Carl to live in his house, and give the botanical lectures for him.

Linnæus was examined by the faculty and judged worthy of being placed (as adjunctus) in Preutz's stead. ' Professor Roberg, however, thought it hazardous to make a teacher of a young man who had not yet been three years a student, and still more so, to entrust him with the public lectures. But there was no other person so proper.[1]

This was in 1730. The young student of twenty-three supplied the aged professor's place with every mark of approbation. The botanical lectures became the talk of Upsala and the attraction of the university. The vivacity of Carl's instructions and the novelty of their matter charmed his audience, accompanied as these were by all the graces of delivery, and the secret of oratory—to be in earnest. His heart was in the work, his handsome face glowing with the love of lovely things as he joyfully taught the students what his superior talent had enabled him to discover. They relished it as our generation has enjoyed receiving light at Ruskin's hands. The effect of his teaching was heightened by the beauty of his voice and diction, and the enthusiasm that fired and enlivened his whole frame, giving a dignity to his personal appearance which had never been remarked before. He seemed born for a professor. The young lecturer himself gained by his residence with Rudbeck an extensive acquaintance with ornithology—a great conquest for one who took the whole of nature for his province; and he now laid the foundation of several of his works—the 'Bibliotheca Botanica,' 'Classes et Genera Plantarum'; for which works Professor Rudbeck's fine collection of books and drawings was of infinite use.

His good fortune did not come single. When one person has made a discovery (of a person, place, or thing)

[1] Diary.

another will usually follow in his wake, the brilliant trail of light being visible and self-evident. Not only was Preutz obliged to give way to Linnæus, who thus, after little over two years' residence at Upsala, was judged qualified to teach the science of botany, but Rudbeck, knowing him to be tutor to Dean Celsius's children, engaged him in the like capacity to his own sons by his second wife. Carl now said grace before full meals; and as the students entertained the most marked contempt for Preutz's abilities, many of them—as Letstrom, Sohlberg, and Archiater Rudbeck's first wife's son, Johan Olof—put themselves under the private instructions of Linnæus. The presents they made him enabled him to assume a more decent appearance in his dress.[1] Dress for gentlemen was a more important and extensive thing than it is now, involving ruffles and embroidery and much fine linen. As adjunctus—oh triumph of all!—he held Rosen's very post.

Now Carl had enough to do; what with the 'Hierobotanicon,' lecturing for Rudbeck, these tutorships, and his private pupils, who flocked to him so soon as he did not need them, and his own books, to say nothing of his researches tacked on to his regular studies in medicine, he was in a whirlwind of work. 'His mornings were passed in giving instruction to pupils and his evenings in composing the new system and meditating a general reform of botanical science. He began his "Bibliotheca Botanica," "Classes Plantarum," "Critica

---

[1] Autobiography.

Botanica," and "Genera Plantarum." Hence, not a moment passed unoccupied during his residence at Upsala.'[1]

But he was so strong and young that nothing came amiss to him. 'Blessed is he who has found his work; let him ask no other blessedness. He has a work, a life purpose; he has found it, and will follow it.'[2] Celsius had brought spring out of the winter of Carl's discontent. Poverty had not narrowed his mind, but now he felt a renewing as he bathed in the bliss of work. It was no fancy, or, if fancy, then 'most real and practical, as many of our fancies are.' He was endowed with twenty-student power—no, twenty-tutor power. No bird, or beast, or insect passed by him unnoticed; while, for the beautiful embroidery of the earth, ah, there are times when, for very gladness, tears only can express our reverence, thankfulness, and perception of the beautiful. Linnæus had little imagination; and if he seemed to lack veneration also, and perception of the beautiful, it was because the artistic capability which expresses these was deficient, yet these things—reverence and perception—were there, unspoken but not unfelt.

Carl now seemed to belong to the successful class, who have never known what it is to lack a meal in their life; and with advantages of dress and pocket-money, he looked a different creature to the lean starving student of last year. Prosperity told upon his humour too: now that he was better off, that he had no

[1] Stoever. [2] Carlyle, *Past and Present.*

gnawings of poverty to contend with, he became popular. The elder professors enjoyed his wit and humour—they love to be lightly amused, the old do; the younger and more earnest sought him for the less sparkling treasures of his mind. A walk with him was of immense interest; he crowded the air and earth with things of life; Pan lived again. He abounded in conversation, and delighted to pour forth the treasures of his knowledge, and thoughts no longer unspeakable; tongue-loosened by the oil and wine of gladness, revealing to their astonishment Nature's open secret. It was not the money he made: it was the fact of success, of appreciation, that made him 'burst out and rollick along in the joy of existence.' Youth had long been stoppered back with him. It was his delight in finding those dreams were true—all with which he used to live in dreamland with Artedi.

Ah, why was there no Boswell at his elbow to colander his best for us? The diaries only give us the bare facts: we know not what it was to hear his thoughts, fresh, full, powerful like a clear mountain stream; but we know for certain it was fine to hear his ideas bubble forth new-born in beauty in his native tongue, for in after years students, ay, and professors, crowded to Upsala simply to hear him speak—in Latin too. These would not have travelled to that far-off nook had not the object been well worth the journey. He was a good listener too, and loved to hear Rudbeck tell all about his journey with his father into Lapland, and the wonders of the great lone North. Eagerly he explored the ruins of old Rudbeck's work and wished the whole could be restored. One of the elder Rudbeck's works he did restore.[1] 'Owing to Rudbeck's age and infirmities the botanic garden had fallen into a very low condition. Carl caused the garden to be entirely altered and planted with the rarest species he could procure, both indigenous and exotic, according to a method of his own. He also instituted botanical excursions with his pupils, who had become numerous. To hear all this must have rejoiced his father. It was, indeed, a hopeful change that Carl was now thought capable of teaching the science of botany, and placed virtually at the head of an establishment in which a year before he had applied for the situation of gardener. Besides botany 'it was decreed,' says Stoever's quaint translator, 'that he should establish a better order in the other reigns of nature, especially among the classes of the animal reign.' One would think he was translating from the French.

We hear less of Artedi now. 'We have been brothers, and henceforth the world will rise between us;'[2] but success did not harden Carl's heart, nor make him unmindful of his beloved friend; while of Celsius he never spoke but in terms of reverence and warmest admiration. Carl possessed greatly the arts of winning and keeping affection.

In Linnæus's eagerness to hear all that the garrulous Rudbeck loved to tell of his father's travels in Lapland, which it had been left him as a sacred legacy to

[1] Diary.   [2] *Paracelsus.*

perpetuate, the professor now thought he saw a way to renew all the parental discoveries. Linnæus's youth and strong constitution, his remarkable powers of mind, and his energy pointed him out as the deputy for the work Rudbeck himself had failed to fulfil. The Royal Academy of Sciences at Upsala had long fostered the hope of forming a complete survey of the whole of Sweden, investigating its capabilities and its natural treasures in order to develop the latent resources of the country. Prompted, doubtless, by Rudbeck's great desire, they proposed to begin by a searching examination of the arctic regions of Sweden. 'They wanted a fresh and virgin intelligence to observe and consider the country.'[1] Celsius and Rudbeck both proposed that Linnæus should undertake the first expedition, and without one thought of the difficulties of the undertaking, the small pay offered, and the disadvantage of being lost to sight of the scientific world for many months, Carl accepted the offer with alacrity, and although the Lapland expedition could not take place till next summer, he at once made his preparations and arranged his affairs, chiefly negotiating the publication of his manuscript books.

The author of 'Spolia Botanica' had not when he wrote it in 1729 'espoused his theory of a sexual difference in the vegetable kingdom, though within three years afterwards it was sufficiently matured in his mind for the arrangement of the Lapland plants in that

[1] D'Israeli's *Endymion*.

method.'[1] Now that he had formed his own system, he seems to have given up all intention of publishing this his earliest work. Linnæus thus advertises the MS. of his second book: 'Upsala, January 1732. A student of medicine and natural history at this university, of the name of Carl Linnæus, takes great pains to represent these two sciences, and botany likewise, in a better light, and to render them more flourishing. The foreign herbs and plants, which are cultivated either in the fields or gardens of Upland, have already been enrolled by him in a little work which appeared last December, 1731, called " Hortus Uplandicus." '

In this book he speaks with praise of his father's garden at Stenbrohult on account of the great number of rare plants in it.

Pulteney, in a footnote, says, 'Stoever mentions a work of Linnæus called " Hortus Uplandicus," which is supposed to be the first in order of time of all his productions; but as the date of it is 1730 [2] it would not have been earlier than the work mentioned above (" Spolia Botanica "). The arrangement is stated to be founded on the doctrine of a sexual difference. I do not find any mention of the " Hortus Uplandicus " in the catalogue of Linnæus's works given in his own diary.' 'Which,' had Stoever but seen the diary, he would comment thus in his polite, roundabout, and long-winded way of learned

---

[1] Pulteney.
[2] This book was written two years before the advertisement of it appeared.

squabble-conducting, 'which by the intensely-respectable-well-born-but-not-his-eyes-using-English-gentleman a careless-and-idiotic-manner-of-precious-and-priceless-documents-an-evidence-of-unenlightenedly-searching-example is.' For the name is in the diary as plain as a pikestaff—and in Pulteney's own book too, second edition. The real difficulty is that he has muddled the dates, both year and month of when the book appeared: errors which Stoever must have crowed over when he met with Pulteney's work.

Linnæus thus prematurely announces another book; or it is advertised for him: 'Upsal, February 15, 1732. An able student of medicine, Mr. Carl Linnæus, causes a botanical work to be printed here, entitled "Fundamenta Botanica."' This did not appear till four years after, in 1736, at Amsterdam. Linnæus sent the MS. to Griefswalda, but he could not find a person who would undertake to publish it. This shows how early Linnæus prepared his system, what alterations he made in the 'Fundamenta Botanica,' and at the same time how eager he was to make his system known, even by advertising works which still remained in MS.

While these things were in preparation who should return but Rosen—return to see his old rival lecturing in his place! One can picture to oneself Linnæus 'biting his lip to keep down a great smile of pride.' How did Rosen like all this? The diary throws some light upon the matter. Late 'in the year 1731, the Medicinæ adjunctus, Dr. Rosen, having returned from his travels abroad, and having perfected himself in anatomy and the practice of medicine, got into universal request, there being no other practitioner at Upsala. He likewise commenced a course of lectures on a branch connected with Professor Rudbeck's office. As the latter was seventy years of age there was a good prospect of his being chosen Rudbeck's successor, and of his having no competitor unless Linnæus got forward. He (Rosen) also applied for permission to lecture publicly on botany, but Rudbeck was unwilling to trust this department to him, as he had never studied it. Rosen tried to persuade Linnæus to give up the lectures to him spontaneously, which Linnæus would have done had Rudbeck consented to it. Thus Linnæus had scarcely surmounted poverty before he became an object of envy—a passion that played him too many tricks, of no use to be mentioned here. The faithless wife of the librarian Norrelius lived at this time in Rudbeck's house, and by her Linnæus was made so odious to his patroness' [Rudbeck's wife] 'that he could no longer stay there; and as Rudbeck had often related to him the curious facts he had noticed and the plants he had discovered on his travels in Lapland, Linnæus conceived a great inclination to visit that country. The secretary of the academy, the Master of Arts, Andres Celsius' [who four years later, 1736, himself visited that country] 'strongly recommended him to go there.'[1] The machinations of his enemies prevailed, and Rosen, who had never been

[1] Diary.

above using mean arts, at length got rid of Linnæus, at any rate for the present. Linnæus left Rudbeck's house and gave up his situation of tutor towards the end of the year, at which time he went to his native province of Småland.[1] Linnæus passed part of the winter with his father at Stenbrohult. A vastly different home-coming from before: the young man honoured with a state commission of importance, the vicarial professor of botany at Upsala, was quite another being from the struggling student who was merely seeking his way. The sisters might well be proud of such a brother, and Gabriel Hök, now rector of the adjoining parish of Wirestad, was glad to make a visit to his old pupil an excuse for also enjoying the society of Linnæus's fair sister. We are not told precisely when the rector of Wirestad married Anna Maria Linnæa, but we may reasonably conclude it was about this time, and it is very probable that Gabriel's bridesman, a clerical friend, on the same happy occasion met and admired the lively and equally pretty sister Juliana, whom he afterwards carried off to another South Swedish rectory.[2]

In January 1732 Carl paid a visit of some days to his kind friend and preceptor Stobæus at Lund, who had by this time forgiven him for leaving his protection. One of Carl's objects in visiting Lund was to study the collection of fossils belonging to Stobæus, this being the only branch of natural history he was not well versed in. Linnæus's mind had grown since he used to look

[1] Diary. [2] Both their portraits are at Hammarby.

at this collection with respectful awe. It is often so with us, the mental garments we once wore with pride no longer fit us. Stobæus's cabinet of minerals, consisting chiefly of petrifactions, did not now satisfy Linnæus.

He returned to his parents' home in Småland, and spent some weeks, and then went to Upsala to prepare for the great journey and learn the result of his publishing negotiations. Just after his return from Stenbrohult another advertisement appears, dated Upsala, March 15, 1732, of the 'Insecta Uplandica,' and a book relating to the birds of Sweden. He took an affectionate leave of Artedi, who was going to England to complete his studies in ichthyology. They made their wills, and the friends mutually assigned to each other such MSS. treating of natural history, as they should be in possession of, in case either of them should die in their travels. Linnæus bore also messages and letters from Artedi to his relatives in Angermania.

CHAPTER VI.

THE NORTH SWEDISH PROVINCES.

Spring clothes the fields and decks the flowery grove,
And all creation glows with life and love.
From the Latin of LINNÆUS.

THE account of Linnæus's Lapland journey was written by himself in a diary called 'Lachesis Lapponica'; this MS. was purchased from the widow of Linnæus, with the rest of the great botanist's writings and collections, by Sir J. E. Smith. It became his duty and wish to render them useful. Great was his disappointment to find the 'Lachesis Lapponica' written in Swedish. For a long time it remained undeciphered. At length Mr. C. Troilus undertook the translation. It proved to be the identical journal written on the spot during the tour; but the difficulty of interpreting it proved unexpectedly great. The bulk of the composition is Swedish, but so intermixed with Latin, even in half sentences, that the translator, not being much acquainted with this language, found it necessary to leave frequent blanks. The translation is in two volumes, octavo.

It is such a journal as a man would write for his own use, without a thought of its ever being seen by any other person. The composition is entirely artless and unaffected, giving a most pleasing idea of the writer's mind and temper, and it is interesting in showing the development of a mind such as that of Linnæus. It is not a professed description of Lapland, nor even a regular detail of the route of the traveller. What was familiar to Linnæus, either in books or in his own mind, he omitted. By the brilliant sketches he has left us in his 'Flora Lapponica,' written in Holland some years later, we see his journal perfected by after-research, which makes it more solid but not so fresh. In the journal we meet with the first traces of ideas, opinions, or discoveries, which scarcely acquired a shape, even in the mind of the writer, till some time afterwards. The familiar and correct use of the Latin language, and the general accuracy of the observations, give a very high idea of the author's accomplishments, considering they are made without a single book to refer to or a companion to consult. The original, moreover, displays a natural eloquence, of which the translation, especially when condensed, falls short. The numerous sketches with a pen that occur in the MS. are strikingly illustrative. His handwriting was small, but legible and elegant.[1] The 'Lachesis Lapponica' had not been translated when Stoever and Pulteney wrote, so that it is here first given with the 'Life.' It is interesting to read this in connection with the journeys of Wheelwright and Du Chaillu on the same roads; Linnæus is

---

[1] Partially abridged from Sir J. Smith's preface.

by far the closest observer of the three, while the difficult Lycksele episode is exclusively his own. His outfit sounds strange 150 years later.

EXTRACTS FROM THE JOURNAL.

' Having been appointed by the Royal Academy of Sciences to travel through Lapland for the purpose of investigating the three kingdoms of nature in that country, I prepared my wearing apparel and other necessaries for the journey as follows.

' My clothes consisted of a light coat of Westgothland linsey-wolsey cloths without folds, lined with red shalloon, having small cuffs and collar of shag; leather breeches; a round wig; a green leather cap, and a pair of half-boots. I carried a small leather bag, half an ell in length, but somewhat less in breadth, furnished on one side with hooks and eyes, so that it could be opened and shut at pleasure. This bag contained one shirt, two pairs of false sleeves; two half-shirts,[1] an inkstand, pencase, microscope, and spying-glass; a gauze cap to protect me occasionally from the gnats, a comb, my journal, and a parcel of paper stitched together for drying plants, both in folio; my MS. " Ornithology," " Flora Uplandica" and " Characteres Generici." I wore a hanger at my side, and carried a small fowling-piece, as well as an octangular stick graduated for the purpose of measuring. My pocket-book contained a passport from the

---

[1] What is a half-shirt?

governor of Upsala, and a recommendation from the Academy.'

I make more copious extracts from the earlier portion of the journal, as the part of Sweden treated of has not often been described. The aspect of the country is almost unaltered since Linnæus's day. In his shorter diary account of his Lapland tour he says he set out on horseback without incumbrances of any kind, and having all his baggage on his back.

'I set out alone from the city of Upsala, on Friday, May 12, 1732 [Old Style], at 11 o'clock, being within half a day of 25 years of age.[1] At this season nature wore her most cheerful and delightful aspect, and Flora celebrated her nuptials with Phœbus.'

A flowery way of saying it was a fine day. Carl seems to have been made vain by the praise bestowed on his eloquence, and to have enriched it at this time by tropes and classical allusions, after the manner of youth. One forgives Linnæus for his flowery language —it was a fashion, like the embroidered waistcoats worn by those dear dandies, the curled darlings of his day—for he wrote sense. It is only 'grand nonsense' that is insupportable.

'Now the winter-corn was half a foot high, and the barley had just shot out its blade. The birch, the elm, and the aspen tree began to put forth their leaves. I left old Upsala on the right, with its three large sepulchral mounds or tumuli. The few plants

[1] A birthday treat of the sort he best enjoyed.

now in flower were *Draba verna*, called in Småland the rye-flower, because as soon as the husbandman sees it in bloom he sows his Lent corn; dandelions, scorpion-grass, violets and wild pansies, *Thlaspi arvense*, *Lithospermum arvense*, sedges,[1] rushes,[2] *Salix*, *Primula veris*, as it is called, though neither here nor in other places the first flower of the spring,[3] the Swedish caper, &c. The lark was my companion all the way, flying before me quavering in the air. *Ecce suum tirile, tirile, suum tirile tractat.*

'Högsta is a Swedish mile and a quarter from Upsala. Here the forests began to thicken. The charming lark here left me, but another bird welcomed my approach to the forest, the redwing, whose warblings from the top of the spruce fir were no less delightful. Its lofty and varied notes rival those of the nightingale herself.'

Linnæus followed the high road, which still exists, like the string of the bow which the railway makes in curving towards Dannemora.

'In the forest are innumerable dwarf firs,[4] whose diminutive height bears no proportion to their thick trunks, their lowermost branches being on a level with the uppermost, and the leading shoot entirely wanting. It seems as if all the branches came from one centre, like those of a palm, and that the top had been cut off.

---

[1] *Carex.*
[2] *Juncus campestris.*
[3] The primrose blooms quite to the end of June in Upland.
[4] *Pinus plicata.*

I attribute this to the soil, and could not but admire it as the pruning of nature.

'At Läby (one and a quarter Swedish mile further) the forest abounds with the Spanish whortleberry, now in blossom. Next came a large and dreary pine forest in which the herbaceous plants seemed almost starved; the soil hardly two inches deep above the sand bore heather, and some lichens of the tribe called coralloides. The Golden Saxifrage [1] was now in blossom.'

He speaks of a runic monument near the posting-house, but the inscription had already been copied.

'Opposite Yfre is a little river, the water of which would at this time have hardly covered the tops of my shoes, though the banks are at least five ells in height. Near the church of Tierp runs a stream whose bank on the side where it curves is very high and steep. The great power of a current, in the way it undermines the ground, is exceeding visible at this place. It now grew late, and I hastened to Mehede, two and a half miles' [S.] 'farther, where I slept.'

He travelled this day over seven and a half Swedish miles, or about fifty English miles.

'May 13. Here the yew grows wild. The forest abounds with yellow anemone, hepatica, and wood-sorrel. Here for the first time I heard the cuckoo.

'Having often been told of the cataract of Elf Carleby, I thought it worth while to go a little out of

[1] *Chrysosplenium alternifolium.*

the way to see it, especially as I could hear its roar from the road, and saw the vapour of its foam rising like the smoke of a chimney. I perceived the river to be divided into three channels by a huge rock. The water in the nearest of these channels falls from a height of twelve or fifteen ells,[1] so that its foam and spray are thrown as high as two ells into the air. On this branch of the cascade stands a sawmill. Below the cataract is a salmon-fishery. Oak trees grow on the summit of the surrounding rocks. At first it seems inconceivable how they should obtain nourishment; but the vapours (of the cataract) are collected by the hills above, and trickle down in streams to their roots. In the valleys I picked up shells remarkable for the acuteness of their spiral points. Here also grew a rare moss of a sulphur-green colour.

'I hastened to the town of Elf-Carleby, which is divided in two parts by the large river. I crossed it by a ferry, where it is about two gun-shots wide. The ferryman' [of course he likens him to Charon] 'asked for my passport, or license to travel. At Elf-Carleby for the first time I beheld what I had never before met with in our northern regions, a peculiar variety of purple anemone [2]—hairy and purplish, stamens numerous and very short.' This flower (a peculiar variety) grows plentifully near Borgholm on the island of Öland. Linnæus also met with it there later.

---

[1] The fall is forty-nine feet high.
[2] *Pulsatilla apiifolia* or *Anemone vernalis*.

'A mile from Elf-Carleby are the iron-works called Härnäs. The ore is brought from Dannemora and from Engsiö in Sudermania. Here runs the river which divides the provinces of Upland and Gestrickland. The post-houses or inns were dreadfully bad. The forests became more hilly and stony, white and dark granite; the rose-willow abounded. Near Gefle stands a runic monumental stone, rather more legible than usual, and on that account better taken care of. By eleven o'clock I arrived at Gefle, where I was obliged to stay all day, for it was evening before I received from the governor of the province of Gestrickland the requisite passport; owing to which delay and my attending morning service next day at Gefle church, I could not quit that place till one o'clock.'

Gefle, with 7,000 inhabitants, is now one of the principal seaport towns of Sweden; well-built and clean, with neat granite quays, and substantial modern appearance.

'At this town is the last apothecary's shop, and also the last physician in the province: these are not to be met with further north. The river is navigable through the town. The surrounding country abounds with large red stones. Here begins a ridge of hills, extending to the next post-house, three-quarters of a mile' [S.] 'further, separating two lakes. In the marshes to the left the note of the snipe was heard continually: on the right are the mineral springs of Hille. Trove post-house, which Professor Rudbeck the elder used to call Troy, is surrounded by a smooth hill. The road from hence lay across a marsh called by the people the walls of Troy. The sweet gale [1] and dwarf birch form a sort of low alley through which the road leads. Here and there grew the marsh-violet with its pale grey flowers, marked with five or seven black forked lines on the lower lip; and in the forests on the other side of the marsh were many kinds of club moss. A quantity of stones lay by the road-side, which the governor of the province had caused to be dug up in order to mend the highway.[2]

'They looked like a mass of ruins, and were clothed with *Campanula serpyllifolia* [the plant afterwards called *Linnæa borealis*],[3] 'whose trailing shoots and verdant leaves were interwoven with those of the ivy. On the right is the lake Hamränge Fjärden, which adds greatly to the beauty of the road. I arrived at Hamränge post-house during the night. The people here talked much of an extraordinary kind of tree: no one could find out what it was. Some said it was an apple-tree which had been cursed by a beggar-woman, who one day having gathered an apple from it, and being on that account seized by the proprietor of the tree, declared that the tree should never bear fruit any more.

[1] *Myrica Gale*, the bog myrtle, or Scotch or Dutch myrtle.
[2] There are four kinds of road in Sweden: the *kungsväg*, king's road, being the finest; country road, *häradsväg* (sometimes called by travellers horrid way), most of which are very good; *sockenväg* (nick-named shocking way), parish road, which is often bad; and the *byräg*, village road, narrow and very rough.—DU CHAILLU.
[3] One of the honeysuckle family.

Next morning I rose with the sun to examine this wonderful tree. It proved to be nothing more than a common elm. Hence, however, we learn that the elm is not a common tree in this part of the country.

'The redwing, cuckoo, black grouse, and mountain finch made a concert in the forest, to which the lowing herds of cattle under the shade of the trees formed a bass. Iceland moss grows abundantly in this forest. I arrived at the river Tonna, which divides Gestrickland from Helsingland and empties itself into the Bay of Touna. The lake called Hamränge Fjärden extends almost to the sea. I was told it did actually communicate' [with the Bothnia]. 'At least there is a ditch in the mountain itself—whether the work of art or nature is uncertain—called the North Sound, hardly wide enough to admit a boat to pass. This is dammed up as summer sets in, to prevent the lake losing too much water by that channel, as the iron from several foundries is conveyed by the navigation through this lake.

'HELSINGLAND.

'The common and spruce firs grow here to a very large size. The inhabitants had stripped almost every tree of its bark. A red byssus stains the stones here, and near Norrala there is a bright red ochre in the earth, and staining the water. Several pairs of semicircular wicker baskets were placed in the water to catch bream. Here I observed the black-throated diver, which uttered a melancholy note, especially in diving.

'In the course of this day's journey I observed a great variety in the face of the country as well as in the soil. Here are mountains, hills, marshes, lakes, forests, clay, sand, and pebbles. Cultivated fields indeed are rare. The greater part of the country consists of uninhabitable mountainous tracts. In the valleys only are to be seen small dwelling-houses, to each of which adjoins a little field. The people seemed somewhat larger in stature than elsewhere, especially the men. The women suckle their children more than twice as long as with us. Brandy is not always to be had here. The people are humane and civilised. Their houses are handsome externally, and neat and comfortable within.' [Are not these advantages due to their having less brandy than elsewhere?]

'The ore used at the capital iron forge of Eksund is of several kinds: first, from Dannemora; second, from Soderom; third, from Grusone, which contains beautiful cubical pyrites; fourth, a black ore from the parish of Arbro, which lies at the bottom of the sea, but in stormy weather is thrown up on the shore.' A kind of blueish stone (*Saxum fornacum*?) is used for building the tunnels and chimneys; it is considered more compact and better able to resist heat than other building-stones. The limestone procured from the seashore abounds with petrified corals.

[1] Of course he calls the workmen sons of Vulcan. Fashions have changed: we never laugh when modern tourists call them sons of Thor, nor when we invoke all the Valkyrie.

'In every river a wheel is placed, contrived to lift up a hammer for the purpose of bruising flax. When it is not wanted, a trap-door is raised to turn the stream aside.

'Several butterflies were to be seen in the forest, as the common black, and the large black and white. Between the post-house of Tygsund and Hudviksvall a violet-coloured clay is found in abundance, forming a regular stratum. I observed it likewise in a hill, which was nine ells in height, near the water, a span-width of violet clay between two layers of barren sand. The clay contained small and delicately smooth white bivalve shells, quite entire, as well as some larger brown ones, of which great quantities are to be found near the water side. At this spot grows the *Anemone hepatica* with a purple flower—a variety so very rare in other places that I should almost be of the opinion of the gardeners who believe that colours of particular earths may be communicated to flowers.

'The produce of the arable land here being but scanty, the inhabitants mix herbs with their corn, and form it into cakes two feet broad, but only a line in thickness, by which means the taste of the herbs is rendered less perceptible. Hudviksvall is a little town situated between a small lake and the sea. Near this place the arctic bramble was beginning to shoot forth, while *Lychnis dioica* and *Arabis Thaliana* were in flower. The larger fields here are sown with flax, which is performed every third year. The soil is turned up by a plough and the seed sown on the furrow, after which the ground is harrowed. The linen manufactory furnishes the principal occupation of the inhabitants of this country. Towards evening I reached Bringstad, and continued my journey at sunrise.

'*May* 17.—I overtook seven Laplanders driving their reindeer, about sixty or seventy in number, followed by their young ones. Most of the herd had lost their horns and new ones were sprouting forth. The drivers spoke good Swedish.

'MEDELPAD.

'Here the common ling grows more scarce, its place being supplied by a greater quantity of the bilberry. Birch trees became more abundant as I advanced. I spied a brace of ptarmigans. All over the country I this day passed the large yellow aconite is as common as ling on a moor. Not being eaten by any kind of cattle, it increases abundantly in proportion as other herbs are devoured. To the north of Dingersjö stands a considerable mountain, called Nyäckersberg, the south side of which is very steep. The inhabitants had planted hop-grounds under it.[1] As the hop does not in general thrive well hereabouts, they designed that this mountain should serve as a wall for the plants to run upon. These hops were very thriving, being sheltered from the north wind and at the same time exposed to the heat of the sun, whose rays are concentrated in this

[1] Ale (*öl*) is the common drink in Sweden and Norway.

spot. I ascended on foot, with a guide, Knorby Kuylen,[1] the highest mountain in Medelpad, finding many uncommon plants in greater perfection than I ever saw them before. The summit is crowned with a beacon used as a signal during the war with the Russians. Every sort of moss grows on this mountain that can be found anywhere in the country round. When at the summit we looked down on the country beneath, varied with plains and cultivated fields, villages, lakes, rivers, &c. Hares run about on the very highest part of this hill. An eagle owl (*Strix Bubo*) rose up suddenly before us as we were sliding down in the descent of the steep south side. We found its nest. Here and there among the rocks were a variety of herbaceous plants, pansies and others. Of the heartsease some of the flowers were white, others blue and white, others with the upper petals blue and yellow, the lateral and lower ones blue, while others again had a mixture of yellow in the side petals. All these were found within a foot of each other; sometimes even on the same stalk different colours were observable—a plain proof that such diversities do not constitute a specific distinction.

'Proceeding farther on my journey, I observed by the road a large reddish stone full of glittering portions of talc. The greater part of my way lay near the seashore, which was strewn with the wrecks of vessels. Towards evening I reached Sundswall, a town situated in a small spot between two high hills. On one side is

[1] Norby Kullen.

the sea, into which a river discharges itself at this place. About sunset I came to Finstad, but continued my route the same evening to Fjähl, where I was obliged to pass a river by two separate ferries, the stream being divided by an island.

'*May* 18.—Being Ascension Day, I spent it at this place, partly on account of the holiday, partly to rest my weary limbs and recruit my strength.

'I was so unfortunate in my journey through Medelpad as not to meet with a single horse that did not tumble with me several times, in consequence of which I was at one time so severely hurt as to be scarcely able to remount. Having already collected a number of stones and minerals, which were no less burdensome than unnecessary to carry with me further, I rode to Hernösand, on the Bothnian Gulf, where I left these encumbrances. I did not, however, stay there above two hours. Near here I picked up a number of chrysomelas' (a sort of beetle) 'of a blueish green and gold.' The city of Hernösand stands upon an island, accessible to ships on every side, except at Värbryggan, where they can scarcely pass.

'I left Fjähl at sunrise, and at Hasjö, the next church, I turned to the left out of the main road to examine a hill where copper ore was said to be found. The stones, indeed, had a glittering appearance like copper ore, but the pyrites to which that was owing were of a yellowish white—a certain indication of their containing chiefly

---

[1] The beautiful *Chrysomela graminis*.

iron. I examined a cave formed by nature in a very hard rocky mountain, formerly a retreat of a criminal who had concealed himself for two years in this retired cavern. The roof and sides of this cave, near the entrance, were clothed with *Byssus cryptarum*. Everywhere near the road lay spar full of talc, or Muscovy glass glittering in the sun. Now we take leave of Medelpad and its sandy roads, as well as its yellow aconite, both of which it possesses in common with Helsingland.

'ANGERMANLAND.

'We no sooner enter this district than we meet with steep and lofty hills scarcely to be descended with safety on horseback.[1] In the heart of the Angermannian forest trees with deciduous leaves, the silver birch, *Betula alba* (with densely matted branches) and the hoary-leaved alder (*Betula incana*) abound equally with the common and spruce firs. These hills might with great advantage be cleared of their wood, for a good soil remains wherever the trees are burnt down—not barren stones as in Helsingland and Medelpad. The valleys between the mountains, as in those countries, are cultivated with corn or laid out in meadows; but here are spacious plains besides. Every house has near

---

[1] On the coast south of Ornsköldsvik the scenery increases in beauty, and as far as Sundsvall the coast is the highest in Sweden. Numerous islands dot the sea along the shore, the principal ones being North and South Ulfö, inhabited by a few hundred fishermen. DU CHAILLU.

it a stage to dry corn and pease on, about eight ells in height, formed of perpendicular posts with transverse beams. The hay, or flax, is hung up to dry on these crossbars of what appears to be a gigantic six-barred gate about twenty feet high. The rye, less plentiful here than barley, is laid here to dry.

'To whatever side I cast my eyes, nothing but lofty blue mountains were to be seen. The little strawberry-leaved bramble (*Rubus arcticus*)[1] was in full bloom. A quarter of a mile further is Doggsta, near which, close to the road, stands the tremendously steep mountain of Skula. This I wished to explore, but the people told me it was impossible. With much difficulty I prevailed on two men to show me the way. We climbed, creeping on our hands and knees, often slipping back again. Sometimes we caught hold of bushes, sometimes of small projecting stones. I was following one of the men in climbing a steep rock, but seeing the other had better success, I endeavoured to overtake him. I had but just left my former situation, when a large mass of rock broke loose from a spot which my late guide had just passed, and fell exactly where I had been, with such force that it struck fire as it went, and was surrounded with fire and smoke. If I had not providentially changed my route nobody would ever have heard of me more. At length, quite spent with

[1] The *Rubus arcticus* is a valuable plant for its fruit, which partakes of the flavour of the raspberry and strawberry, and makes a most delicious wine, used only by the nobility in Sweden.—SMITH.

toil, we reached the object of our pursuit, which is a cavity in the middle of the mountain—a mere cavern. The stones that compose it are of very hard quality, or spar; yet the sides of the cavern are in many places as even as if they had been cut artificially. Several different strata are distinguishable, particularly in the roof, which is concave like an arch. In that part a hole appears, intended, I was told, for a chimney. Several sorts of ferns grow on the adjacent parts of the mountain. We descended with much greater ease. Laying hold of the tops of spruce firs which grew close to the rocks, we slid down upon them, dragging them after us down the precipices.

'I had scarcely continued my journey a quarter of a mile before I found a great part of the country covered with snow, in patches some inches deep. The pretty spring flowers had gradually disappeared. The buds of the birch, which so greatly contribute to the beauty of the forests, were not yet put forth. The high mountains which surround this track and screen it from the genial southern and western breezes may account for the long duration of the snow.

'The cornfields afford a crop two years successively, and lie fallow the third. Rye is seldom or never sown here, being too slow in coming to perfection; so that the land, which must next receive the barley, would be too much exhausted.

'*May* 21.—After going to church at Natra I remarked some cornfields, which the curate had caused to be cultivated in a manner that appeared extraordinary to me. After the field has lain fallow three or four years it is sown with one part rye and two parts barley mixed together. The seed is sown in spring, as soon as the earth is capable of tillage. The barley grows rank, ripens its ears, and is reaped. The rye meanwhile goes into leaf, but shoots up no stem, as the barley smothers it and retards its growth. After the latter is reaped the rye advances in growth, and ripens the year following without any further cultivation, the crop being very abundant. The inhabitants here also make broad thin cakes of bread. The flour used for this purpose commonly consists of one part of barley and three of chaff. When they wish to have it very good and the country is rich in barley, they add but two portions of chaff to one of corn. The cakes are not suffered to remain long in the oven, but require to be turned once. Only one is baked at a time, and the fire is swept towards the sides of the oven with a large bunch of cock's feathers. The coverlets of the beds at this place are made of hare-skins. To-day I met with no flowers except the wood-sorrel, which here is the primula, or first flower of spring. The lily of the valley and strawberry-leaved bramble were plentifully in leaf.

'*May* 22.—Apple trees grow between Veda and Hornoen, but none are to be seen further north. No kind of willow is to be met with throughout Angermanland, nor is the hazel. Cherries do not always ripen, but potatoes thrive very well. Tobacco and hops both grow slowly and are of rare occurrence.' [One marvels that tobacco grows at all. It shows that the sun shines down very hot there, and brings an annual plant quickly forward.] 'In the road I saw a cuckoo fed by a *Motacilla* (water wagtail?). Near the coast was a quicksand, caused here, as in Skåne, by the fine light sand of the soil being taken up by the wind into the air and then spread about upon the grass, which it destroys. The road in several parts lies close to the seashore.

'*May* 23.—After having spent the night at Normaling, I took a walk to examine the neighbourhood, and met with a mineral spring, already observed by Mr. Peter Artedi at this his native place. It appeared to contain a great quantity of ochre, but seemed by the taste too astringent to be wholesome.

'I observed on the adjacent shore that an additional quantity of sand is thrown up every year by the sea, which thus makes a rampart against its own encroachments, continually adding by little and little to the continent.[1]

'In proportion as I approached Westbothland, the height of the mountains, the quantity of large stones, and the extent of the forests gradually decreased. Firtrees, which of late had been of rare occurrence, became more abundant.

---

[1] Angermanland is a beautiful province, and many of its valleys are very productive. The Angermanelfven, running through its whole territory, is the deepest river of Sweden, and may be ascended by steamboats as far as Nyland, sixty [English] miles, and by small craft to Holm thirty miles farther.' The river is two miles wide at Wedga beyond Hernösand.—DU CHAILLU.

'WESTBOTHLAND.

'The ground here is tolerably level; the soil sand, or sometimes clay. In some places are large tracts of moss. Thus the country is by no means fertile, though it affords a good deal of milk. Barley is the chief grain raised here. No flowers were to be seen here—not even the wood sorrel, my only consolation in Angermanland. The two sorts of cotton-rush were now coming into bloom. The dwarf birch was abundant enough, but as yet showed no signs of catkins or leaves. Throughout the whole of this country no ash, maple, lime, elm, nor willow is to be seen, much less hazel, oak, or beech. Towards evening I reached Röbäck, where I passed the night.

'*May* 24.—Close to Röbäck is a fine spacious meadow, which would be quite level were it not for the hundreds of ant-hills scattered near it. Near the road, and very near the rivulet that takes its course towards the town of Umeå, are some mineral springs, abounding with ochre, and covered with a silvery pellicle. I conceive that Röbäck may have obtained its name from this red sediment—from *röd*, red, and *bäck*, a rivulet.'[1]

Carl was ferried over to Umeå[2] by a 'brawny bald grey-headed, grey-coated Charon,' just such as Rudbeck had described to him.

[1] Not a difficult guess.
[2] 'Umeå,' a little dirty old town, with a remarkably fine white church, and the largest prison I have seen in the North.—WHEELWRIGHT.

'Baron Grundell, the governor of the province, a pattern of mildness, received me kindly, showed me several curiosities, and gave me much interesting information. The birds I saw here were the crossbill (which cleverly fed on the cones of the spruce fir), yellowhammers, swallows, snow-buntings and ortolans.

'Ruffs and reeves had been in plenty this year. In the cornfields lay hundreds of gulls (*Larus canus*) of sky-blue colour.

'In the garden the governor' [the pattern of mildness] 'showed me orach, salad, and red cabbage,[1] which last thrives very well, though the white cabbage will not come to perfection here; also garden and winter cresses, scurvy-grass, camomile, radishes, goosetongue (*Achillea ptarmica*), rose-campion, wild-rose, lovage, spinach, onions, leeks, chives, cucumbers, columbines, carnations, sweet-william, gooseberries, currants, the barberry, elder, guelder-rose, and lilac. Potatoes here are not larger than poppy-heads; tobacco, managed with the greatest care, and when the season is remarkably favourable, sometimes perfects seed. Dwarf French beans thrive pretty well, but the climbing kinds never succeed. Broad beans come to perfection; but peas, though they form pods, never ripen. Roses, apples, pears, and plums hardly grow at all, though cultivated with the greatest attention. Cherries, apples, pears, and plums always fail.[2]

---

[1] If he had a good cook for these herbs, this is, perhaps, the origin of his designation.

[2] In Umeå Du Chaillu saw a garden filled with flowers, straw-

The people wear a kind of shoes, or half boots, called *kängor*, easy in wearing and impenetrable to water. Those who walk there may walk in water up to the tops without wetting their feet, for the seams never give way as in our common shoes. They only cost two copper dollars. They are cut so that not a morsel of leather is wasted. Thick soles are here needless; neither are heels wanted. Nature, whom no artist has yet been able to excel, has not given (high) heels to mankind, and for this reason we see the people of Westbothland trip along as easily and nimbly in these shoes as if they went barefoot.

'*May* 26.—I took leave of Umeå and turned out of the main road to the left, my design being to visit Lycksele Lapmark. By this means I missed the advantage I had hitherto had at the regular post-houses, of commanding a horse whenever I pleased, which is no small advantage to a stranger travelling in Sweden. It now became necessary for me to entreat in the most submissive manner when I stood in need of this useful animal. The road grew more and more narrow and bad, so that my horse went stumbling along at almost every step among great stones at the hazard of my life. My path was so narrow and intricate along so many by-ways that nothing human could have followed my track. In this dreary wilderness I began to feel very solitary

berries, raspberries, currant bushes, peas, carrots, and potatoes, with a stretch of green fields beyond. Cauliflowers, cabbage, and lettuce had headed, peas were bearing fully, and melons were growing under glass.

and to long earnestly for a companion. The mere exercise of a trotting-horse in a good road, to set the heart and spirits at liberty, would have been preferable to the slow and tedious mode of travelling which I was doomed to experience. The few inhabitants I met with had a foreign accent, and always concluded their sentences with an adjective. Here grew a willow [1] very hairy all over; its catkins were for the most part advanced and faded.

'In the evening I arrived at Jamtboht, where some women were sitting employed in cutting the bark of the aspen tree into small pieces scarcely an inch long and not half so broad. The bark is stript from the tree just when the leaves begin to sprout, and laid up in a place under the roof of a house till autumn or the following spring, when it is cut up to serve as good for cows, goats, and sheep, instead of hay, a very scarce article in these parts, for the fields consist principally of marshy tracts with coarse herbage. On my inquiring what I could have for supper they set before me the breast of a cock of the wood (*Tetrao Urogallus*), which had been shot and dressed some time the preceding year. Its aspect was not inviting, and I imagined the flavour would be not much better, but I was mistaken. The taste proved delicious, and I wondered at the ignorance of those who, having more fowls than they know how to dispose of, suffer many of them to be quite spoiled, as often happens at Stockholm. After the breast is

[1] *Salix lanata.*

plucked, separated from the other parts of the bird, and cleaned, a gash is cut longitudinally on each side of the breast-bone quite through to the bottom, and two others parallel to it a little farther off, so that the inside of the flesh is laid open in order that it may be thoroughly dressed. The whole is first salted with fine salt for several days. Afterwards a small quantity of flour is strewed on the under side to prevent its sticking, and then it is put into an oven to be gradually dried. When done it is hung up in the roof of the house, to be kept till wanted, where it would continue perfectly good even for three years if it were necessary to preserve it so long.

'It rained so violently that I could not continue my journey that evening, and was therefore obliged to pass the night at this place. The pillows of my bed were stuffed with reindeer's hair instead of feathers. Under the sheet was the hide of a reindeer with the hair on, the hairy side uppermost, on which people told me I should lie very soft. They use willow bark for tanning the leather.

'*May* 27.—At noon I pursued the same bad road as yesterday—the worst road I ever saw, made of stones piled on stones among large entangled roots of trees. The frost, which had just left the ground, made matters worse. All the elements were against me. The branches of the trees hung down before my eyes, loaded with raindrops in every direction. Wherever any young birches appeared they were bent down to the earth' [across the path], 'so that it was difficult to pass them. The aged pines, which for so many seasons had raised their proud tops above the rest of the forest, overthrown by the wrath of Juno (!), lay prostrate in my way. The rivulets, which traversed the country in various directions, were very deep, and the bridges over them so decayed and ruinous that it was at the peril of one's neck to pass them on a stumbling horse.

Many persons had confidently assured me that it was absolutely impossible to travel to Lycksele in the summer season; but I had always comforted myself with the saying of Solomon (?) that 'nothing is impossible under the sun.' However, I found that if patience be requisite anywhere, it is in this place. To complete my distress, I had a horse whose saddle was not stuffed, and instead of a bridle I had only a rope, which was tied to the animal's under jaw. Here and there in the heart of the forest were level heathy spots, as even as if they had been made so by a line, consisting of barren sand, on which grew a few straggling firs and some scattered plants of ling. Some places afforded the perforated coralline lichen (*L. uncialis*), which the inhabitants in rainy weather, when it is tough, rake together in large heaps and carry home for the winter provender of their cattle. These sandy spots, about a mile' [Swedish] 'in extent, were encompassed as it were with a rampart or very steep bank fifteen or twenty ells in height, so nearly perpendicular that it could not be ascended or descended without extreme difficulty.

It often happened that above one of these sandy heaths lay another equally barren. The interstices of the country between these embanked heaths were occupied by water, rocks, and marshes, producing abundance of firs, intermixed with some birches, all covered with black and white filamentous lichens. The few small juniper bushes were all close pressed to the ground. At Abacken, and on the road beyond it for a considerable way, some loose ice still remained, which surprised me much at this season of the year;[1] yet I recollected I had but a week before met with snow near Mount Skula.

'Nothing but water can be had to drink. Against the walls of the houses an agaric, shaped like a horse's hoof,[2] was hung up to serve as a pincushion. As a protection against rain the people wear a broad horizontal collar made of birch bark, fastened round the neck with pins.

'The women wash their houses with a kind of brush made of twigs of spruce fir which they tie to the right foot and scrub the floor with it. The peasants, instead of tobacco, smoke the buds of hops, or sometimes juniper berries or the juniper bark.

'In the evening I reached Texnäs in the parish of Umeå. Seven miles' [Swedish] 'distant from this place is the church, the road to which is execrable, so that the people are obliged to set out on Friday morning to get to church on Sunday. On this account they can

---

[1] This is June in the New Style.   [2] *Boletus igniarius.*

seldom attend Divine service, except on fast days, and Whitsunday, Easter, and Christmas days. Timber for the purpose of building a church here was brought so long ago as the time of the late Abraham Lindelius;[1] but it has lain till it is rotten.

'*May* 28.—I left Texnäs and proceeded to Genom, where I was obliged to stay till next day, as there is no conveyance but by water to Lycksele, and the wind blew very hard.' Here he saw a beaver, which he describes.

'*May* 29.—Very early in the morning I quitted Genom in a *haep*, or small boat, proceeding along the western branch of the Umeå River. When the sun rose nothing could be more pleasant than the view of this clear unruffled stream, neither contaminated by floods nor *disturbed by the breath of Æolus*. All along its translucent margin the forests which dotted its banks were reflected like another landscape in the water. On both sides were large level heaths guarded by steep ramparts towards the river, and these were embellished with plants and bushes, the whole reversed in the water, appearing to great advantage. The huge pines, which had hitherto braved Neptune's power, smiled with a fictitious shadow in the stream. Neptune, however, in alliance with Æolus, had already triumphed over many of their companions: the former by attacking their roots, while the latter had demolished their branches.

'Close to the shore were many ringed plovers and sandpipers.' [He saw also owls, white swans, and

[1] Was this an ancestor of Linnæus?

cranes.] 'The peasant who was my rower and companion had placed about thirty small nets along the shore, in which he caught pike. A dried pike of 20 lbs. weight is sold for a dollar and five marks, silver coin. In one of the nets he found a large male goosander caught.

'The river along which we had rowed for nearly three miles' [S.], 'and which had hitherto been easily navigable, now threatened us with interruptions from small shelves forming cascades, and at length we came to three of these, very near each other, which were absolutely impassable. One of them is called the waterfall of Tuken. My companion, after committing all my property to my care, laid his knapsack on his back and turning the boat bottom upwards, placed the two oars longitudinally, so as to cross the seats. These rested on his arms as he carried his boat over his head, and thus he scampered away over hills and valleys, so that the Devil himself could not have come up with him.' Linnæus made a sketch of the boat, which was in 'length 12 feet, breadth 5 feet, depth 2 feet. The four planks which formed each of its sides were of root of spruce fir; the two transverse seats were of branches of the same tree; the seams were secured obliquely with cord as thick as a goose-quill.' He gives a humorous sketch of the man running off with the boat, half covered with it. 'Now and then some poplars are to be seen. The forest was rendered pleasant by the tender leaves of the birch, more advanced than any I had hitherto met with. Among the plants were golden rod, marsh marigold, and the *Linnæa borealis*; among birds, the ringed plover, the redwing, the tufted duck, and the black-throated diver. A little before we reached the church of Lycksele, a fourth waterfall presented itself. This is more considerable than the preceding, and falls over a rock. On its brink the curate had erected a mill. Some islands of considerable size are seen in the river as we approach this waterfall. The adjoining mountain is formed of a mixed spar, and extends a good way to the right, being in one part very lofty, and perpendicular, like a vast wall, towards the shore. At eight in the evening I arrived at the hospitable dwelling of Mr. Oladron,[1] the curate of Lycksele, who, as well as his wife, received me with great kindness. They at first advised me to stay with them till the next fast day, the Laplanders not being implicitly to be trusted, and presenting their firearms at any stranger who comes upon them unawares or without some recommendation. In the morning (May 30), however, my hosts changed their opinion, being apprehensive of my journey being impeded by floods if I delayed it.'

Here he gives drawings and descriptions of the paraphernalia used in driving the reindeer; the ornaments of the saddlery, harness, and so forth. 'The pasture-ground near the parsonage of Lycksele was very poor, but quite the reverse about quarter of a mile distant.

[1] Or Pastor Gran. In the abridged account of his tour, drawn up as a report to the Academy, this name is given. Possibly the name was Olaf Gran.

Here the butter was remarkable for its fine yellow colour, approaching almost to a reddish or saffron hue; wherever the birch abounded the pasture-ground was of the best quality. In the school here were only eight scholars. The church was in a miserable state.

'At Whitsuntide this year no Laplanders were at church, the pike happening to spawn just at that time. This fishery constitutes the chief trade of these people, and they were therefore, now, for the most part dispersed among the Alps, each in his own tract, in pursuit of this object. Divine service being over, on May 31 I left Lycksele in order to proceed towards Sorsele.'

In this tour he describes the *Linnæa borealis*. His own 'neglected fate and early maturity are said to be typified by it.' He gathered it at Lycksele on May 29, and chose it for his own especial flower. Hitherto this elegant and singular little plant had been called *Campanula serpyllifolia*, thyme-leaved bell-flower; but Linnæus, prosecuting the study of vegetables on his new principle,[1] soon found this to constitute a new genus. He reserved the idea, keeping it warm in his heart, till his discoveries and publications had entitled him to botanical commemoration, and his friend Gronovius, in due time, with his concurrence, undertook to make this genus known to the world. It was published by Linnæus himself in the 'Genera Plantarum,' 1737, and in the same year in the 'Flora Lapponica,' with a plate.

[1] Smith.

It is mentioned in the 'Critica Botanica' as 'an humble, despised, and neglected Lapland plant, flowering at an early age.' This he regarded as typical of himself.

Linnæa borealis grows in shady places in Scotland, Switzerland, Canada, &c., and was cultivated (after Linnæus became famous) in the Jardin du Roi in Paris. The plant has a slight perfume in the evening. It is said to be specific against gout and rheumatism; though Linnæus, who suffered from these complaints, never mentions the plant as medicinal.

Fries[1] speaks of the Linnæa as 'one of the prettiest of plants, which by its colours and its exquisite vanilla perfume enlivens the dark pine woods of Sweden.'

At this place too, Lycksele, he seems to have adopted the motto *Tantus amor florum*, 'Thus great is the love of flowers.'

[1] The present Director of the Botanical Garden at Upsala.

CHAPTER VII.

'LACHESIS LAPPONICA': JOURNEY THROUGH LAPLAND, 1731, MAY TO NOVEMBER.

> Men there are whose patient minds,
>   In one object centred,
> Wait, till through their darkened blinds
>   Truth has burst and entered.
> Then, that ray so barely caught
>   Joyfully absorbing,
> They behold the realms of Thought
>   Into Science orbing.
>
> . . . . . .
>
> Men there are whose ambient souls,
>   In rapt Intuition,
> Seize Creation as it rolls,
>   Whole, without partition.—J. C. MAXWELL.

'WE here behold, not the awful preceptor of the learned world in his professorial chair, but a youthful inexperienced student full of ardour and curiosity, such as we ourselves have been.'[1] This Lapland journey was the first and most difficult of the six travels of Linnæus: a sort of labours of Hercules. Even now the young inquirer asks concerning Lapland: 'Haven't you got to eat bears' grease there always?' To know the country better is to find that there is very little bears'

---

[1] Sir J. E. Smith.

grease to be had. A hundred and fifty years ago, when Linnæus travelled, the country was not known at all; Rudbeck's memorials were destroyed and his son's memory was failing. The utmost that was known of Lapland had been learnt by Linnæus sitting at the feet of the younger Rudbeck before his memory failed him altogether. It was a Robinson-Crusoe-like form of journey; for not only did Carl travel alone, but he met with the scantiest of population, in miles and miles of loneliness studded with here and there a cottage. Excepting in the larger towns and on board the steamers, the population of Sweden is still everywhere 'understood but not exprest.' Here in Lapmark it is not even understood: the country is one vast emptiness, like the rest of the world in the days of Paradise; peopled only by the 'lovely phantoms of the waterfalls.'

The intrepid hardy-bred Linnæus, with his untiring energy, was the very man to undertake a journey of discovery like this. He observed everything: had an eager appetite for all forms of nature. His indomitable industry was well suited to that interminable Lapland day 'in which one loses all hope that the stars and quiet will ever come.' It only enlarged his opportunities to see 'the dawn shine through the whole night till it be morning.' To be out and away into the wide open, was his longing desire. He had studied books enough; now for the mind's liberty, now to range through broad nature. To educate is to set free the mind, new sculptured, from its marble block. Truly this journey was broad enough. Round Lapland, skirting the boundaries of Norway, he returned to Upsala by the eastern side of the Bothnian Gulf, having in five months travelled nearly 4,000 English miles, much of it on foot. That many modern travellers and sportsmen do the same is only to say that many people go to America and many view the Pacific 'from a peak in Darien'; but for all that there is but one Columbus and one Cortez. Linnæus's journey is as good as a guide-book even now, for the face of the country is unchanged, and he is as clearly descriptive as Baedeker or Murray. Even Du Chaillu scarcely reads clearer, fuller, or more modern. I select such portions from the two volumes as best illustrate his character and history.

'*May* 31.—The Divine service of this day being over I left Lycksele for Sorsele, taking with me only three loaves of bread and some reindeer tongues by way of provision. I presumed that I should procure among the Laplanders reindeer-flesh, cheese, milk, fish, fowl, &c. Nor indeed could I well take anything more at present; for whenever we came at any shoals or falls in the river my companion took our boat on his head over mountains and valleys, so that I had not only my own luggage to carry but my guide's likewise. At one place, close to the river, was a Laplander's shop raised on a round pole as high as a tall man and as thick as one's arm. This pole supported a horizontal beam, with two cross-pieces, which together formed the foundation of the edifice. The walls are very thin; the ceiling is of birch bark, with a roof of wood and stone above it. It is scarcely possible to conceive how the owner can creep into this building, the door being so small, and wherein he is like a bird in a tree. The birch bark is extremely useful to the Laplanders: they make their plates or trenchers of it, and boat-scoops, shoes, tubs to salt fish in, and baskets. They also tan their leather with birch bark, like the Russians.[1]

'*June* 1.—We pursued our journey by water with considerable labour and difficulty all night long—if it might be called night, which was as light as day, the sun disappearing for about half an hour only, and the temperature of the air being rather cold. Fir trees were thinly scattered, but they were extremely lofty. Here were spacious tracts producing the finest timber I ever beheld. The ground was covered with ling, red whortleberries,[2] and mosses. In the low grounds grew smaller firs, amongst abundance of birch, and red whortleberries, which grew larger as he travelled northward, as well as the common black kind.[3] On the dry hills, which most abounded with large pines, the finest timber was strewed around, felled by the force of the tempests. The Laplanders formed their huts of these. The huts were at this time mostly deserted. We found guides in various Laplanders, and proceeded up the

---

[1] The oil from this bark gives the peculiar odour to Russia leather.
[2] *Vaccinium Vitis Idæa.* Idæan vine, as Scott called it in the 'Lady of the Lake.' Idæan, relating to Mount Ida in Crete.
[3] *Vaccinium Myrtillus.*

Umeå River, turning off to the right at the Juita branch. Here I found crake-berries,[1] as large as the black bilberry, and herb Paris. But what most surprised and pleased me was the little round-leaved yellow violet[2] described by Morrison, which had not before been observed in Sweden.

'I shall not dwell on the inconveniences I had to undergo every time we had to seek for any of the Laplanders, while I was quite destitute of provisions. These poor people themselves had at this season nothing but fish to eat, as they had not yet begun to slaughter their reindeer nor to go a-fowling; neither had they as yet milked any of their reindeer.

'*June* 2.—We were obliged to leave our boat; the river being so rapid, and so much impeded by falls, that we were obliged to undertake a walk of a few miles' [Swedish] 'further, which I was told would bring us to a more navigable stream. A fen or marsh lay before us, seemingly half a mile' [Swedish] 'broad, which we had to to cross. At every step the water was above our knees, and ice was at the bottom. Where the frost was quite gone we often sunk still deeper, sometimes to the waist. If we thought to find footing on some grassy tuft it proved treacherous and only sunk us lower. Sometimes we came where no bottom was to be felt, and had to measure back our weary steps. Our half-boots were filled with the coldest water. When we had traversed this marsh we sought in vain

---

[1] *Empetrum nigrum.*   [2] *Viola biflora.*

for any human creature, and were therefore under the necessity, a little further on, of crossing (in pursuit of my new Lapland guide) another bog still worse than the former, and a mile' [Swedish] 'in extent. I know not what I would not rather have undertaken than to pass this place, especially as it blew and rained violently. We reposed ourselves about six in the morning, wrung the water out of our clothes, while the cold north wind parched us as much on one side as the fire we lighted scorched us on the other, and the gnats kept inflicting their stings. I had now my fill of travelling. These marshes are called *stygx*. The Styx of the poets could not exceed them in horror. We now directed our steps to the desert of Lapmark, not knowing where we went' [in the diary account of his tour he calls this place Olycksmyran—the unlucky marsh]. 'My Laplander, after a weary search, brought a woman of very diminutive stature to see me, who addressed me in Swedish in the following terms: "O thou poor man! what hard destiny can have brought thee hither, to a place never visited by anyone before? This is the first time I ever beheld a stranger. Thou miserable creature! How didst thou come, and whither wilt thou go?" I inquired how far it was to Sorsele. "That we do not know," replied she, "but in the present state of the roads it is at least seven days' journey from hence, as my husband has told me.' There was no boat to be had on the next river. It was not possible to proceed further in this direction, and we had to return by the horrible way we came. The good woman conducted us to a side path, whereby we avoided about half a mile' [Swedish] 'of the way we had come. In a shed supported by four posts hung some clothes and a small reindeer cheese, which I wished to purchase. The woman refused, as she wanted it herself; but my hunger was such that I could not lose sight of this cheese. "I have no desire," she said, "that thou shouldst die in my country for want of food," and at last she let me buy it.' Even she was struck with his wretched appearance.[1] 'We continued our voyage down the river, being carried with great velocity by the current, the whole of the next day. At length coming to an island, the Laplander failed in his attempt to weather it, and the boat, striking against a rock, was dashed to pieces. We both found ourselves in the water. My conductor lost not only his boat, but a hatchet and a pike. I lost two stuffed birds—one a large heron, black, with a white breast; the other a red bird, or *gvousach*, as the Laplanders call it.[2] With difficulty we got from this island to the shore.[3] The sun shone warm, and after having wrung the water out of our clothes we walked on for about a mile' [Swedish] 'along the bank of the river, amongst thickets and bogs, till we came in sight of a colonist who was fishing for

[1] 'He actilly looked as if had been picked off a rock at sea and dragged through a gimlet-hole.'—S. SLICK.

[2] *Corvus infaustus.*

[3] He thinks first of the loss of the birds; his own rescue is a minor detail.

pike. He gave me some provision, and conducted me to Grano, where I only stopped to rest one night, and on the evening of June 8 arrived at Umeå. These poor people roast their fish thoroughly, and boil it better and longer than ever I saw practised before. They know no other soup or spoon-meat than the water in which their fish has been boiled. I could not observe that the nights were at all less light than the days, except when the sun was clouded. On the banks of the river, where fragments are to be found of all the productions of the mountains, I met with silver ore.

'A Laplander, whose family consists of four persons, including himself, when he has no other meat, kills a reindeer every week, three of which are equal to an ox; he consequently consumes about thirty of those animals in the course of the winter, which are equal to ten oxen, whereas a single ox is sufficient for a Swedish peasant. The bountiful provision of nature is evinced in providing mankind with bed and bedding even in this savage wilderness. The great hair moss [1] is used for this purpose. They choose the starry-headed plants, out of the tufts of which they cut a surface as large as they please for a bed and bolster, separating it from the earth beneath. This mossy cushion is very soft and elastic, not growing hard by pressure; and if a similar portion of it be made to serve as a coverlet, nothing can be more warm and comfortable. They fold this bed together,

---

[1] *Polytrichum commune.*

tying it up into a roll that may be grasped by a man's arms, which, if necessary, they carry with them to the place where they mean to sleep the night following. If it becomes too dry and compressed, its elasticity is restored by a little moisture.

'*June* 12.—I took my departure (from Umeå) very early in the misty morning. The sun appeared quite dim, wading, as it were, through the clouds. *Andromeda polifolia* was at that time in its highest beauty, decorating the marshy grounds. The flowers are quite blood-red before they expand, but when full-grown the corolla is of a flesh-colour. Scarcely any painter's art can so happily imitate the beauty of a fine female complexion;[1] still less could any artificial colour upon the face itself bear a comparison with this lovely blossom. As I contemplated it I could not help thinking of Andromeda as described by the poets, which seemed so applicable to the plant before me, that if these writers had had it in view they could scarcely have conceived a more apposite fable. This plant is always fixed on some turfy hillock in the midst of the swamps, as Andromeda herself was chained to a rock in the sea which bathed her feet, as the fresh water does the roots of the plants; dragons and venomous serpents surrounded her, as toads and other reptiles frequent the abode of her vegetable prototype, and, when they pair in the spring, throw mud and water over its leaves and branches. As the distressed virgin cast down her blush-

---

[1] A Swede can judge of fine complexions.

ing face through excessive affliction, so does the rosy-coloured flower hang its head, growing paler and paler until it withers away. Hence, as this plant forms a new genus, I have chosen for it the name of Andromeda.[1]

'All the woods and copses by the way abounded with butterflies of the fritillary tribe without silver spots. An elegant little blackish butterfly, besprinkled with snow-white spots like rings, smooth and lustrous on the under side, was very plentiful in the paths. The great dragon-fly, with two flat lobes at its tail, and another species with blue wings, were also common.

'The poorer Laplanders rock their infants on branches of trees.[2] In the part of the country where I was now travelling the cradles rock vertically, or from head to foot.

'I now entered the territory of Piteå. Here I met with kind entertainment from Mr. Solander, the principal clergyman of the place.'[3] [He shot and sketched a *Strix ulula*, which was too much damaged to allow of stuffing.] 'Just at sunset on June 15 I reached the town of Old Piteå, having crossed the broad river in a ferry boat. Immediately on entering the town I procured a lodging, but had not been long in bed before I perceived a glare of light on the wall of

---

[1] Linnæus has carried the fanciful analogy farther in his *Flora Lapponica*: 'At length comes Perseus in the shape of Summer, dries up the surrounding water, and destroys the monsters, rendering the damsel a fruitful mother, who then carries her head (the capsule) erect.'

[2] 'Hushaby, baby, on the tree-top.'

[3] Father of Dr. Solander the naturalist.

my chamber. I was alarmed with the idea of fire, but on looking out of the window saw the sun rising, perfectly red, which I did not expect would take place so soon. The cock crowed, the birds began to sing, and sleep was banished from my eyelids. Near the new town of Piteå, close to the shore, grew the round-leaved water-violet,[1] with perfectly snow-white flowers.

'*June* 19.—I went out to sea in a boat for some miles' [Swedish] 'to explore the neighbouring coast and islands, and returned at length to the new town. In the island of Longoen, three miles' [S] 'from Old Piteå, I was lucky enough to find growing under a spruce fir the coral-rooted orchis (*Ophrys corallorrhiza*) in full bloom. It is a very rare plant. I proceeded to Luleå, being desirous of reaching the alps of Lulean Lapland in time enough to see the midnight sun, which is seen to greater advantage there than at Torneå. The new town of Luleå is very small, situated on a peninsula encompassed by a kind of bay. The soil is barren. Indeed the slight eminence the town stands on is a mere heap of stones, with sea-sand in their interstices. It seems as if the sea had carried away all the earth, and, like a beast of prey, had left nothing but the bones, throwing sand over them to conceal its ravages. As no horse was to be procured in the whole place, I proceeded by sea to Old Luleå, half a mile' [Swedish] 'distant. Here the curious kind of grass [2] which is called in Småland "old man's beard" is known by the name of *Lapp-här*, "Lap-

---

[1] *Viola palustris.*   [2] *Nardus stricta.*

lander's hair." It was now in blossom. There is great conformity between this country and Småland.' His old home friend the bird-cherry (*hägg*) grew plentifully here—it grows even within the arctic circle. 'Many herbaceous plants grow here which are not to be found in Upland, Sudermania, Ostrogothia,[1] nor Skåne, though natives of Småland. The water swarmed with innumerable fishes just spawned, so pellucid that they were rendered conspicuous chiefly by their large eyes. The observer of nature sees, with admiration, that the whole world is full of the glory of God.'[2]

The weather had become fair, and Linnæus says, 'If the summer be indeed shorter here than in any other part of the world, it must be allowed at the same time to be nowhere more delightful. I was never in my life in better health than at present.'

'*June* 24 (*Midsummer Day*) [July 4 N. S.].—Blessed be the Lord for the beauty of summer and of spring, and for what is here in greater perfection than almost anywhere else in the world—the air, the water, the verdure of the herbage, and the song of birds!'[3]

'*Sunday, June* 20.—After Divine service I took leave

[1] East Gothland.

[2] The parish church of Luleå is regarded as the oldest in Westbothnia, having been built in the very earliest ages of Christianity, and was very famous while the Catholic religion prevailed in Sweden. It contains a remarkable old altar-piece, the gilding of which cost 2,408 ducats. In the vestry a copy of the canonical law, in seven vols. folio, is still preserved.

[3] Du Chaillu says: 'I had not before heard so many birds singing together after midnight, enjoying the spring. Before two o'clock the swallows were out of their nests.'

of Luleå. Half-way between Svarlå and Harns I met with the (*Pedicularis*) *Sceptrum Carolinum*, first observed by Professor Rudbeck. This stately plant was not yet in flower. It grew in a dry soil. Near Harns is found a fine handsome blue clay, in some measure fireproof; also a rare kind of iron ore.' He notes the purple *Pinguicula*,[1] June 29, in Luleå-Lapland—and 'a *Pinguicula* the fore-part of whose petal was white, the hind part blue, which is certainly a beautiful as well as singular variety. The little alpine variety of the ptarmigan[2] was now accompanied by its young. I caught one of these, upon which the hen ran so close to me that I could easily have taken her also. She kept continually jumping round and round me; but I thought it a pity to deprive the tender brood of their mother; neither would my compassion for the mother allow me long to detain her offspring, which I restored to her in safety.

'I embarked on the Luleå River, which I continued to navigate upwards for several days and nights, having good accommodation both as to food and boat. After three days and nights we reached Quickjock.

'My companion was a Laplander, who served me both as servant and interpreter. Few persons are met with on these alps who speak Swedish, and I had already suffered much in the Lapland part of Umeå for want of knowing the language. Nor was a companion less required to assist me in carrying

[1] Butterwort. [2] *Tetrao Lagopus*.

what was necessary, for I had sufficient encumbrances of my own without being the bearer of our provisions into the bargain. The pine trees are more barren of branches on their north sides; hence the people know by these trees which way the north lies. Brandy is here made from the fir, as well as from the berries of the mountain ash. The *Angelica sylvestris* is a dainty in great request among the Lapps; they use its root for the cure of their terrible colic. The common method of the Laplanders for joining broken earthenware is to tie the fragments together with a thread and boil the whole in fresh milk, by which they are cemented to each other. The reindeer milk is very glutinous.

'*July* 1.—When I came to the lake Skalk in the way towards Kionitis I was much struck with an opening between the hills to the N.W., through which appeared a range of mountains, from ten to twenty miles' [Swedish] 'distant, as white as the clouds, and seeming not above a mile' [S] 'from the spot where I stood. Their summits reached the clouds, and indeed they resembled a range of white clouds rising from the horizon. They recalled to my mind the frontispiece of Rudbeck's "Lapponia Illustrata." Mountains upon mountains rose before me in every direction. In one word, I now beheld the Lapland Alps.' [He gives a clever pen-and-ink sketch of the view.] 'At Kionitis I rested during the whole of Sunday, July 2. Here the beautiful corn was growing in perfection in valleys between the snowy mountains. It had shot up so high as to be laid in some places by the rain. It was sown on May 25 or 26, as at Umeå.

'"Alone I crossed the Hyperborean tracts of ice, the snowy Tanais, and fields never free from Riphæan frosts."'—*Virgil*, 'Georgics,' iv. 517. [Linnæus quotes the passage.]

'After several days' travelling, on the evening of July 6 I ascended Vallivari, the first mountain of the Alps on this side. On my first ascending these wild alps I felt as if in a new world. I saw few birds, except some ptarmigans running with their young along the vales. The declining sun never disappeared sufficiently to allow any cooling shade. The midnight sun, deep red, glowing like a fierce charcoal fire, tinged everything with roseate hue, most magical upon the snow, bewildering the brain, and producing a drowsy effect.[1]

The peak of Sulitelma is 6,326 feet high. The blue glaciers hereabout are magnificent. Du Chaillu says he never lost sight of the blue outline of Sulitelma, but the peak was mostly hidden from view.[2] Linnæus specifies no peak as Sulitelma, but only speaks of the mass as Vallivari. When at length he was able to turn

---

[1] Often was I seized with an indescribable feeling of loneliness, and at the same time a desire to wander farther away.—DU CHAILLU.

[2] As the sun shone upon the ice its hue was simply marvellous; it seemed in many places like a huge mass of sparkling topaz; its extent was enormous, and patches of snow were scattered over its surface. There were only two breaks of dark rock visible in the frozen mass; and towering above all was Sulitelma, dark and gloomy, looking down upon the sea of ice.—DU CHAILLU.

his eyes from the magic of the mountains Linnæus was equally enchanted with the new world of arctic nature at his feet.

'When I cast my eyes over the grass and herbage there were few objects I had seen before, so that all nature was alike strange to me. I walked in snow as if it had been the severest winter. All the rare plants that I had previously met with, and which had from time to time afforded me so much pleasure, were here as in miniature, and now also in such profusion that I was overcome with astonishment, thinking I had now found more than I should know what to do with. I sat down to collect and describe these vegetable rarities.'

He gives a list of thirty, all described and named extemporaneously.[1] Not one of these names has subsequently been set aside by any of his severest critics. He noted the silken-leaved alpine lady's mantle, the deep green sibbaldia, the little purple-flowered azalea, the succulent rose-root, the red lychnis, and several ranunculi, the beautiful saxifraga stellaris, rivularis, and oppositifolia, of which last Du Chaillu says, 'Many times have I remained standing in admiration before this exquisite flower, which looks like a velvety carpet of purple moss, and grows in patches on the dark rocks, often surrounded by snow.' And the *Primula farinosa* of which Linnæus speaks,—
'This primula, the splendid crimson of whose flowers attracts the eyes of all who traverse the fields of Skåne

---

[1] One plant was dedicated subsequently to Jussieu.

and the meadows of Upland in early spring, did not occur during my whole journey till after I had ascended the Lapland Alps'—here it was as the face of a friend.

'The time passed unperceived away, and my interpreter was obliged to remind me that we had still five or six miles' [Swedish] 'to go to the nearest Laplander, and that if we had a mind for any reindeer meat we ought to bestir ourselves quickly.' They had been refreshed by the snow-water running down in streams. They hastened on and reached the summit of the ridge, standing on the brow of Vallivari, ' from hence the verdant appearance of Norway, lying far beneath us, was very delightful. The whole country was perfectly green, and, notwithstanding its vast extent, looked like a garden in miniature, for the tallest trees appeared not above a span high. Our calculations were very inadequate to what we found its actual distance. At length, however, we reached the plains of which we had enjoyed so stupendous a prospect. Nothing could be more delightful to my feelings than this transition from all the severity of winter to the warmth and beauty of summer. The verdant herbage, the sweet-scented clover, the tall grass, reaching up to my arms, the grateful flavour of the wild fruits, and the fine weather which welcomed me to the foot of the alps, refreshed me both in mind and body.'

'NORWAY.

'At the place where I stopped to rest after my fatiguing journey they gave me sword-fish [1] to eat, which much resembled salmon in flavour.

'Here I found myself close to the sea-coast. I took up my abode at the house of a shipmaster, with whom I made an agreement to be taken in a boat the following day along the coast. I much wished to approach the celebrated whirlpool called the Maelstrom, but I could find nobody willing to venture near it. We set sail next morning, according to appointment, but the wind proved contrary, and the boatmen were after a while exhausted with rowing. Meanwhile I amused myself in examining various petrifactions, principally medusæ, zoophytes, and submarine plants of the Fucus tribe, which occupied every part of the coast. I was kindly received at the house of the pastor of Torfiorden, who had an extremely beautiful daughter, Sarah Rask, eighteen years of age. I must not omit to write to him hereafter; for, according to his account, he never expected to see an honest Swede.' [By the Norwegians the Swedes were always accounted fair and false, as Scott says of the Scots.] 'Next day we proceeded further on our voyage and returned to our place of departure, the wind being still contrary.' [They could get no further than Rörstad church, near the mouth of the fiord.] 'On the following morning I climbed one of the neighbouring mountains with the intention

[1] *Xiphias gladius.*

of measuring its height. While I was reposing tranquilly on the side of the hill, busied only in loosening a stone which I wanted to examine, I heard the report of a gun at a small distance below. I was too far off to receive any hurt, but perceiving the man who had fired the gun, I pursued him to a considerable distance in order to prevent his charging his piece a second time. I could get no explanation of this attack.

'I saw no flies in Lapland, but in Norway the houses are full of them. I was, however, no longer infested with swarms of gnats.[1]

'On July 15 we set out on our return, and that whole day was employed in climbing the mountains again, the ground being extremely steep as well as lofty. It is customary for those in our part of Sweden who fancy themselves indisposed to frequent watering-places or mineral springs during the heat of summer. For my own part I have, thank God, for several years enjoyed tolerable health ; but, as soon as I got upon the Alps I seemed to have acquired a new existence. I felt as if relieved from a heavy burthen ; and after having spent a few days in the low country of Norway, though without having committed the least excess, I found my languor or heaviness return. When I again ascended the Alps I revived as before. The Lapland water, too,

---

[1] Du Chaillu says 'until the end of June there are no gnats.' I suspect they are local, for I have seen them in swarms on May 29, and I have passed in previous and following weeks through many provinces of Sweden without seeing any gnats at all.

is uncommonly grateful to the palate. Since I set out on my journey I have become able to walk four times as far as I could at first, yet I could not but wonder at my two Laplanders—one of them upwards of seventy—who had accompanied me during the whole of this day's tedious walk. While I was resting they played and frisked about. This set me seriously to consider the question put to me by Dr. Rosen: "Why are the Laplanders so swift-footed?" To which I answer that it arises not from any one cause, but from the co-operation of many. 1. They wear no heels to their half-boots' [2, 3, 4, and 5 are also very good reasons, but foreign to our present purpose].

'Every Laplander constantly carries a sort of pole, tipped with a ferule, and furnished with a transverse bar of wood. When he is tired he leans his arms and nose against it to rest himself.' Linnæus gives a drawing of their snowshoes, as also of their chessboard, and describes at length the rules of their elaborate game. Their chess king has a castle and eight Swedes his subjects; sixteen Muscovites are their adversaries. Several games are common among these people, who, for all their hard climate and circumstances, are by no means always at work.

'We turned our course towards the alps of Torneå, which were described to me as about forty (Swedish) miles distant' [270 English miles]. 'What I endured in the course of this journey is hardly to be described. How many weary steps was I obliged to set to climb the precipices that came in my way! Sometimes we were enveloped with clouds' [the greatest danger in Lapland travel, because of the unseen precipices], 'sometimes rivers impeded our course and obliged us to choose a very circuitous path, or to wade naked through the snow-cold water.

'Without the fresh snow-water, our only drink, we should never have been able to encounter the excessive heat of the weather.[1]

'Having nearly reached the Lapland village of Caituma, the inhabitants of which seemed perfectly wild, running away from their huts as soon as they perceived us approaching from a considerable distance, I began to be tired of advancing further up into this inhospitable country. We had not tasted bread for several days, our stock being exhausted, and the rich milk of the reindeer is too luscious to be eaten without bread. I was desirous of having my linen washed, but the people understood my request as little as if I had spoken Hebrew, not a single article of their own apparel being made of linen.

'The dwarf birch bears very small leaves in those

---

[1] Du Chaillu records his experience on the same line of travel:—
'Of all the bleak landscapes I had seen on the journey this seemed the most dreary; it was absolutely grand in its desolation. There was an indescribable charm in the loneliness and utter silence; bare mountains of granite and gneiss formed the setting of the picture, and all around were stones of all sizes and shapes, piled in heaps. Over these we had to wind our way for hours, jumping from one to another almost continuously. All the hard pedestrian exercise I had ever taken was as nothing compared to this.'

elevated regions. In this part of the country the crakeberry (*Empetrum*) serves for firing; otherwise the most common fuel is the dwarf birch and the willow,[1] with white hairy leaves, so abundant on the Lapland Alps.'

In the 'Flora Lapponica' he describes one of these night journeys, when the low-focussed central light was sending the long oblique shadows in dense blue bands round a crimson world. 'While I was walking quickly along over the celebrated mountain of Vallivari, facing the cold wind at midnight—if I may call it night when the sun was shining without setting at all—I perceived, as it were, the shadow of this plant (*Andromeda tetragona*), but did not stop to examine it, taking it for the *Empetrum*. But after going a few steps farther, an idea of its being something I was unacquainted with came across my mind, and I turned back, when I should again have taken it for the *Empetrum* had not its greater height caused me to consider it with more attention. I know not what it is that so deceives the sight in our Alps during the night as to render objects far less distinct than in the middle of the day, although the sun shines equally bright. The sun, being near the horizon, spreads its rays in such a horizontal direction that a hat can scarcely protect our eyes; besides, the shadows of the plants are so infinitely extended, and so confounded with each other from the tremulous agitation caused by the blustering wind, that objects very different in

---

[1] *Salix Lapponum.*

themselves are scarcely to be distinguished from each other. Having gathered one of these plants, I looked about and found several more in the neighbourhood, all on the north side, where they grew in plenty; but I never met with the same in any other place afterwards. As at this time they had lost their flowers, and were ripening seed, it was not till after I had sought for a very long time that I met with a single flower, which was white, shaped like a lily of the valley, but with five sharper divisions.'

'*July* 24.—This night I beheld a star, for the first time since I came within the arctic circle. Still I could see to read or write easily enough.'[1]

Linnæus now determined to return towards Quickjock, a journey of about forty Swedish miles. In the course of this journey he met with an accident which might have been serious. Walking over the snow, he broke through the icy crust covering a deep hole. This cavity was very steep, and so hollowed out by the water that it surrounded our traveller like a wall. The guides could not release him until they had procured a rope, when he was drawn out, with no other injury than a hurt on the thigh, which continued to be felt for a month afterwards.

'*July* 25.—The lakes in this part of the country

[1] Du Chaillu travelling here about the same time in July and August says, 'I was gladdened by the view of a star, the first I had seen for about three months. It was *Vega*, twinkling bright, an old friend, who had often helped me to find my way through the African jungle.'

did not afford me so many plants as those further south. Their bottoms were quite clear and destitute of vegetation. The shores were no less barren. No water-lilies, no water-docks, &c., grew about their borders, but the surface of the water itself was covered with the water ranunculus, bearing round as well as capillary leaves, and whitening the whole with its blossoms. I could not but marvel to see these broad patches of white spread over the lakes, as when I passed up the country only a fortnight before I had not perceived the least appearance of even the herbage of the ranunculus that composed them; now its branches, an ell in length, swam on the surface. The growth of the stem must be very rapid, as it often proceeded from a depth of three fathoms.

'At sunset we reached Parkjaur, where we vainly attempted to procure a boat. We had no resource but to make ourselves a float or raft, on which we committed our persons and our property to the guidance of the current of the river. The night proved very dark—in consequence of a thick fog, insomuch that we could not see before us to the distance of three fathoms. After a while we found ourselves in the middle of the stream, and it was not long before the force of the water separated the timbers of our raft, and we were in imminent danger of our lives. At length, however, with the greatest difficulty, we reached a house situated on an island, after a voyage of half a mile' [Swedish] 'from where we had embarked.

'The next day I was conducted to the river of Calatz, to see the manner of fishing for pearls,' from the then nearly exhausted bed of pearl-mussels. He carried thence the germ of an idea of pearl-making with him, brooding over it for years.

'*July* 28.—Several days ago the forests had been set on fire by lightning, and the flames raged at this time with great violence, owing to the drought of the season. In many different cases, perhaps in nine or ten that came under my notice, the devastation extended several miles' [Swedish] ' distance. I traversed a space three quarters of a mile ' [Swedish] ' in extent which was entirely burnt ; so that Flora, instead of appearing in her gay and verdant attire, was in deep sable—a spectacle more abhorrent to my feelings than to see her clad in the white livery of winter. The fire was nearly extinguished in most of the spots we visited, except in ant-hills and dry trunks of trees. After we had travelled about half a quarter of a mile across one of these scenes of desolation the wind began to blow with more force than it had done, upon which a sudden noise arose in the half-burnt forest, such as I can only compare to what may be imagined among a large army attacked by an enemy. We knew not whither to turn our steps. The smoke would not suffer us to remain where we were, neither durst we turn back. It seemed best to hasten forward, in hopes of speedily reaching the outskirts of the wood; but in this we were disappointed. We ran as fast as we could in order to avoid being crushed by the falling trees, some of which threatened us every minute. Sometimes the fall of a huge trunk was so sudden that we stood aghast, not knowing whither to turn to escape destruction, throwing ourselves entirely on the protection of Providence. In one instance a large tree fell exactly between me and my guide.

'This day I observed the harvest beginning. The corn now cutting at Torneå, though sown but a few days before midsummer, was nevertheless quite ripe.

'On July 30 I arrived at Luleå. I visited the Laxholms, islands so called from the salmon-fishery.[1] Those who fish for salmon come to this place about a fortnight before midsummer, and remain till St. Bartholomew's Day, August 28, as during that space of time the salmon keep ascending the river. Few of the fish escape being taken so as to return down the river. At Michaelmas the fishermen come here again, when they catch a smaller sort of salmon.

'I rested for a day or two, and then proceeded to Torneå.

'*August* 3.—At sunrise the marshes were all white with hoar-frost. In the preceding night winter had paid his first visit and slept in the lap of the lovely Flora.

'On leaving Sangis I left my mother-tongue behind me. At Saris I met with native Finlanders only, whose language was unintelligible to me. Between this and Torneå are three ferries to pass.

[1] *Salmo salar*, named *Lax* by the Swedes.

'*August* 7.—The town of Torneå stands on a small island—I call it an island because it is bounded on the north by a swamp, on the south-east by the great river of Torneå, and on the west and south-west by a shallow arm of the sea. No kind of plough is used at Torneå, the ground being turned over with the spade.'

Linnæus detected the cause of a dreadful disease among the reindeer of North Lapland: some had died in the winter, but more in the spring when turned out to grass. He discovered the water-hemlock,[1] one of the most virulent of vegetable poisons, growing in the marshes. By pointing out the plant he enabled the people to guard against the danger ever after. He recommended the Torneans to employ people to root it out.

'Not understanding the Finnish language, I found it inconvenient to proceed, and preferred returning. I made several excursions to an adjacent island.

'*September* 4.—I went to Biorknäs in order to be instructed in the art of assaying. Here I stood sponsor to the son of the burgomaster (or mine-master) Swanberg, who was born in the preceding night.' In the summary of his travels he mentions how, on his return through Luleå, he learned the art of assaying from the mine-master Swanberg, at Calix, in two days and a night; and having suffered extreme fatigue, he reposed himself at the house of M. Hoyer, the magistrate.

'*September* 14.—I took my leave of Biorknäs. The

[1] *Cicuta virosa.*

weather was cold and rainy. Such of the forest trees as are of a deciduous nature had now assumed a pallid hue in consequence of the cold nights, but the evergreens' [that is, the pines] 'were rendered conspicuous by their dark green colour. The hills appeared sandy, and such places as had been burnt were now perfectly white with reindeer moss.

'*September* 15.—I received one hundred dollars, of copper money, from the chief clergyman at Torneå.' [This seems to have been left here in deposit for him by the Academy.]

'Having noted the Finnish names for such articles as I should be most likely to want at the inns, I ventured once more to enter East Bothland, in order to pursue my journey that way homeward. I considered that in a new country there is always something new to be seen, and that to travel the same road I had come would probably afford but little entertainment or instruction. I had still less inclination, at this advanced season of the year, to encounter the hazard of a sea voyage. I therefore pursued my way along the coast through East Bothland and Finland, visiting Uleå, Brakestad, Old and New Carleby—the latter is as big as Wexio—Wasa, a handsome little town, the residence of the governor, Christinestad, Biorreberg, and Åbo, seat of the Finland university, remaining four days at the place last mentioned. I then went by the post-yacht to Åland, crossed the Sea of Åland, and at one in the afternoon on October 10 arrived safe at Upsala. To the Maker and Preserver of all things be praise, honour, and glory for ever!

'The whole extent of my journey amounts to 633 Swedish miles' [about 3,798 English miles].

Linnæus speaks very modestly of this journey in his diary. 'On his arrival at home he delivered to the Academy of Sciences an account of his expedition, which obtained their approbation, and they returned him 112 silver dollars (not more than 10l. sterling),[1] his travelling expenses. They also elected him one of their members.' He considered his labour amply repaid by the payment of his expenses, the information he had gained, and the discovery of new plants upon the higher mountains. He has eulogised the country in the 'Flora Lapponica' as all that could be desired; happy and smiling, free from many diseases and the scourge of war, and possessing plentiful resources in itself; while the inhabitants are said to be innocent and primitive, displaying the greatest hospitality and kindness to a stranger.[2] 'See what pure nature could do for these men,' cries Linnæus; but this was the memory of a Swede in Holland. The journal shows us the seamy side.

It is amusing to read in Smith's preface, 'So valuable was the MS. of the Lapland tour considered, that on Linnæus's whole collection and library being sold,

---

[1] Pulteney. Mr. Jackson, Secretary to the Linnæan Society, reckons these 112 dollars as less than 25l. sterling.

[2] Sir W. Jardine.

after the death of his son, it was remarked that these papers at least ought to have been retained in Sweden as a national property, the journey which they record having been undertaken *at the public expense*, and the objects illustrated thereby being necessarily more important to the author's countrymen than to any other people.'

CHAPTER VIII.

ROSEN VICTOR.

> I am as earnest as a bee,
> But savage as a hornet.—FARRAR.

AFTER his tour. Linnæus again felt the pressure of poverty, as one cannot live only upon fame. Immediately after his return from Lapland he made application for Wrede's exhibition, called *Öfverskotts medlen*, which he obtained chiefly by the kind assistance of Professor Valraves. From this he enjoyed the first year 30 plåtar (about 5*l*. sterling). I can discover no other university prize obtained by Linnæus while an undergraduate.

He was no longer tutor to Rudbeck's sons, nor could he live with Rudbeck as before, on account of the aforementioned feminine influence. But Menander, afterwards Bishop of Åbo, was at that time a student, and assisted Linnæus considerably with money: the latter taught him natural history in return.

Having learnt the art of assaying metals during his ten days' residence at the mines of Biorknäs, near Calix, in the course of his Lapland tour, Linnæus, early in

1733, began a private course of lectures on this subject. The novelty of his information, the vivacity of his style, and the grace of his delivery soon gained him celebrity in this line also. Linnæus had a general elegance of manners in common with most Swedes; but beyond this, as was said of our Dr. Johnson, 'few persons quitted his company without perceiving themselves wiser and better than they were before'; while, as a lecturer, he had the faculty of expressing what he meant to convey in clear incisive words, in sentences vigorous and full, from his complete mastery of the subject. One relished hearing him as one enjoys seeing a master workman use his tools.

He was above his age in the same sense that the flower is above the plant, that the sunflower crowns the stem. In him the natural arrogance of youth was not the arrogance of a fool swollen with conceit and vapour, but the arrogance of Aristotle's 'man of lofty soul,'[1] who, being of great merit, knows that he is so and chooses to be so regarded.'[2] He had passions—' passions in general lofty and generous, but still passions.' Though entirely free from malice, he was impulsive and vehement in temper, and when roused to indignation could be very fierce.

Few persons have all kinds of merit belonging to their character; 'a fallible being will fail somewhere: as

[1] Froude.

[2] 'It is the heroic arrogance of some old Scandinavian conqueror; it is his nature and the untamable impulse that has given him power to crush the dragons.'—M. FULLER, speaking of Carlyle.

Johnson says, 'It is well where a man possesses any strong positive excellence.' Rosen, his old rival, whose position as adjunctus to the chair of Anatomy and Physics [1] gave him great weight in the university, owing to Professor Roberg's advanced age and weakness, had his envy roused by Linnæus's rising fame; he was not above taking mean measures to rob the brilliant young lecturer of his reputation, and even threatened to stop his lectures as illegal. All this made Linnæus very bitter. 'There is no precedent for this, as I am the first person who has ever lectured in this way,' said Linnæus when one day he called on Rosen, hoping to settle the matter in talk. Rosen, sitting grumpy as a polar bear, eyed him with suspicion and distrust, 'and would not come forth into open parley at all.'

'There is a rule against such lectures,' said Rosen. 'An obsolete regulation,' retorted Linnæus. The constant opposition his natural bent met with on all hands had doubtless its result in deepening in Linnæus a certain irascibility of temper that often underlies the sweetness of the Swedes. The old Goth peeps out, the Berserk spirit of the Saga heroes.

That he in the fulness of his strength should not be let to use it were unreasonable and unnatural. Work had to be done, ideas to be enlarged: was he not to be permitted to do these things and to maintain himself? It was a manifest injustice, under which he could not but smart. His clever tongue had a sting in it too, as

[1] Physiology and physics were formerly considered as synonymous.

we can tell from a letter of Haller's. 'The man is active, I cannot deny, and a zealous lover of nature, for which I love him; but his character has for me a something—I know not what to call it—of asperity, fickleness, and unevenness.' The fact was, his vanity clashed against their vanity. Unless very first-class men themselves, they were afraid to measure tongues with him; they were fearful, too, lest he should spy out and expose the poverty of their land.

Linnæus shall now state his own case in an extract from his diary.

'In the year 1733 Linnæus began a course of lectures in the art of assaying, which had never been before taught in this university. He delivered them for 2 plåtar' [about 7s.] 'each person, on which account he gained a great number of pupils. Rosen, observing that Linnæus came forward more and more, and fearing lest he should at last become a dangerous competitor, requested Linnæus to lend him his MS. lectures on botany, which he had himself composed, and which he valued more than anything that belonged to him; and when Rosen found he could not attain them by fair means he held out threats to Linnæus, who then gave up to him a part of them; but as soon as he was informed that Rosen copied the MS. no intimidation could induce him to deliver into Rosen's hands the remainder. In the meantime Rosen had taken by the hand a young Master of Arts, named Gottskalk Wallerius, who had studied medicine under him almost a year. The office of adjunctus in the medical faculty at Lund was now instituted, and Linnæus endeavoured to obtain it at the urgent desire of Professor Rudbeck. Rosen was at this time (1733) practising at Wiksberg, where people went to drink the mineral waters. The chancellor of the university, Count Carl Gyllenborg, was of the number, and consequently Linnæus stood no chance against Wallerius, who obtained the office of adjunctus, though it was of less advantage to him than it would have been to Linnæus.'

Disappointed in his views of medical advancement, Linnæus turned his attention to mineralogy, one of the kingdoms of his universal empire.[1]

Being prohibited from publicly lecturing, Linnæus accepted the invitation of some of his former pupils to accompany them to the mines of Falun and other places.

'On his return from Lapland Linnæus paid particular attention to mineralogy, which was the principal reason of his visiting the district of mines—a spot the most favourable of all others for acquiring that knowledge of minerals which could alone enable him to form a correct system.'[2] He was impressed, besides, with the interdependence of the natural sciences.

At the end of the year 1733 Linnæus went to the mine district—called in Sweden Bergslag, in a dreary desolate country resembling the bleak high Cornish

---

[1] In 1733 he studied mineralogy and the docimastic art.—*Encyl. Brit.*, eighth edition.

[2] Diary.

mining districts—for the purpose of investigating and arranging the minerals of his native country, where he visited Norberg, Bispberg, Afvestad, Garpenberg, a sort of quadrilateral of mines, and the iron-foundries, mines, and town of Falun, which place he has memorialised by his *Lichen Faluniensis*, a production more resembling some ramifications of the neighbouring copper ores than anything of vegetable origin. Linnæus was received in this rich but desolate mining district with the most flattering distinction, and attentions were paid him which were heard of in Upsala. He was introduced to Baron Reuterholm, the governor of the province, who requested Linnæus to undertake, at his (Reuterholm's) expense, a journey all over Dalecarlia, with other naturalists, to survey the physical productions of that province.[1]

Reuterholm delighted in the study of nature, and chiefly spent his leisure hours with the productions of the mines. His charge as director of the mines became more lucrative in proportion to his knowledge of their produce. He also wished his sons to learn these things, and he rejoiced in their rapid liking for the gifted young stranger, and encouraged their intimacy. Baron Reuterholm was himself charmed with the enthusiastic young man who descended the mines by day and passed the night in the foundries by the furnaces. So practical, too, he was, that, not satisfied with discovery, he at once sought to put every material to use. 'It is not the

[1] Diary.

finding of a thing, but the making something out of it after it is found, that is of consequence.'[1]

Reuterholm meant to put Linnæus to use likewise. He persuaded him to undertake the travelling tutorship of his two sons through Dalecarlia and over the Dalecarlian Alps to Norway—an idea which soon after enlarged itself into having a complete survey of the province, in every department of its natural history, undertaken at the governor's expense—a patriotic work, by which also Baron Reuterholm expected largely to profit. But nothing was definitively settled about the undertaking, and the idea went to sleep during the night of the Swedish winter.

Linnæus remained a whole month at Falun, and then returned to Upsala.

'The very distinction he had so justly acquired turned out to his prejudice. Envy and rivalship combined with self-interest gave rise to all the violence of animosity. Linnæus had not taken his degree, which according to the Swedish custom must always be taken abroad, and Linnæus was too poor to travel. This excluded him from the right of delivering public lectures, which is the exclusive privilege of doctors. He was too obnoxious to his competitors, who were determined to check his rising fame.'[2]

The applause which he received was unendurable to Rosen, now (in 1734) become a more formidable enemy through his marriage with the niece of the Archbishop of Upsala.[3] Rosen, conceiving that the genius and reputa-

[1] J. R. Lowell. [2] Stoever. [3] Turton.

tion of Linnæus stood in the way of his own fame, and attracted to the new doctrines some of Rosen's own pupils, determined to suppress his competitor. Sir J. E. Smith, who also speaks strongly on the mean jealousy of Rosen in surreptitiously copying Linnæus's botanical manuscripts, of which he had the forced loan, mentions as 'the basest action of Rosen, and which proved envy to be the sole source of his conduct, this, that, having married the niece of the archbishop, he obtained through his lordship's means an order from the chancellor to prohibit all private medical lectures in the university. This, for which there could be no motives but conscious inferiority and malice, deprived Linnæus of his only means of subsistence, and the students of any information which might endanger their reverence for his rival.' Smith is very bitter on the prosperous nephew of an archbishop.

'Rosen procured an edict from the Chancellor Cronhjelm, that a medical teacher should never be received in the university of Upsala to the prejudice of the adjunctus.'[1]

There was no precedent, it would seem, concerning private lectures. Linnæus in his tabular summary says: '1733. Lectured privately on mineralogy—he was the first person who had done so—at Upsala.'[2] The regulation concerning public lectures, it appears, existed before, but had fallen into abeyance or been forgotten through there being no outsider competent to lecture, as we hear that Rosen informed against his rival 'before

[1] Diary. [2] Notes made by Linnæus for his biography.

the senate of the university, and insisted that in virtue of the academical statute Linnæus should be no longer suffered to give public lectures.'[1] He thus meanly sought to strangle the reputation of Linnæus and deprive the world of the benefit of his knowledge because it was not sanctioned by academic forms.

The proud spirit of Linnæus had to submit to all the vexations and restrictions entailed on him by his poverty —griefs more galling, perhaps, to bear than were his actual hardships when an undistinguished student. Poverty is a mighty strengthener as well as tamer and chastiser; but for this discipline at Upsala Linnæus would probably never have vanquished the world with his system. He had now nothing but private lectures to depend upon. 'Ah!' cries Dante (in the 'Paradiso'), 'if the world but knew the heart of him who goes from trouble to trouble, begging his life!'

Linnæus was summoned to appear before the senate. Many of the members were anxious to waive the prohibition in consideration of the virtues and talents of him at whom it was now pointed;[2] but Rosen pleaded the inviolability of the statutes, which the senate was bound to enforce, and Linnæus was forbidden to continue his lectures. Rosen was prepared with his special edict, pointed directly against Linnæus and his lectures, public or private, in case of the votes going against him.

This was a dreadful blow to Carl. His ambition

[1] Stoever. [2] Ibid.

hemmed in the sphere of its operations, no outlook was open to him. 'The bitterest of griefs is to know much and accomplish nothing.'[1] Linnæus was terribly sore, and no wonder. It was his ruin. 'When roused I am like a furious bard of ancient days. I poured forth such a dreadful torrent of sarcasm and truth that I shook him to death,' says our English painter,[2] when likewise chafing under ill-treatment.

No wonder if the wrath of Linnæus burst forth in a most unbounded manner. The wild-beast vein of the ancient Goth rose in him. In the tempest of his passion he forgot himself, his future happiness, and every moral consideration, but ' who ever saw far in a storm ' ? Boiling with pugnacity and rage, with flaming eyes more piercing than his knife, he swore ' By all the Valkyrs ! ' he would slay his foe.

When Rosen left the senate Linnæus waited for him, and with desperate fury drew his sword, and would have run it through the body of his enemy had not the bystanders fortunately wrested it from him. He flew at Rosen's throat and grappled with him in a fierce struggle. He was with difficulty separated from his prey.[3] Rosen, who was a member of the academy, complained of this gross assault and of this daring violation of the laws of public safety. The rigour of the law threatened Linnæus with proscription, and he could never afterwards have made his appearance at Upsala. Dean Celsius interposed, allayed the resentment caused by this event,

---

[1] Herodotus.   [2] Haydon.   [3] Stoever.

and brought round matters so far that punishment was changed into a bare reprimand. Linnæus was now spared penalty, but he still cherished the idea of vengeance. His fiery temper almost drove him to desperation. Rising to a white heat, he still meditated the design of stabbing Rosen if he met him in the streets.

I can realise the transports of fury of Linnæus because I have twice seen men of these Northern nations give way to fits of frenzy of the like sort. Once it was a gentleman who had quarrelled with some of his countrymen in a railway-carriage about a mere trifle. Rendered speechless by his own fury, all trembling with passion, he became an amusing spectacle of pantomimic rage. The other case was a more serious affair, and likely to become tragic. It was a common man in the island of Gothland, who, in an outburst of savage wrath only comparable to that of an ancient Berserk, hurled the huge stones that lay about the cliff in his madness at his enemy, who fled terrified into a house near by. A woman came out and faced the seeming maniac, crossing her arms proudly as if she said, 'You pass this threshold only over my body.' I must do the furious Goth the justice to say that in his wildest transports he hurled no stone against the woman—a noble-looking creature—though at every minute the demon repossessed him; and he flung his body and the great boulders about blindly and with renewed vehemence as the thought of his wrongs rushed over him again. Such was Linnæus at this time; a renewed personal struggle with Rosen would have been like Molin's fine statue of the belt-wrestlers [1] at Stockholm, which represents the jealous wrestlers struggling to the death with their sharp short knives, such as are used among these Northmen to this day, and bound together by a strong leathern belt in order that the fight may only end with the death of one or both of them.

But that history says it was a sword, I should think it more probable that Linnæus rushed on Rosen with the stout sharp two-edged weapon, the tolle-knife, that the Scandinavians so generally wear in an ornamented sheath at their thigh; though, as the trial was a ceremonial occasion, dress swords may have been worn.

Duelling, to which severe penalties were attached by a law of 1682, had long been unknown in Sweden. It was advantageously replaced in the universities by 'national'[2] quartette-singing, in which Linnæus seldom or never joined.

Thus cruelly deprived of resources which promised an ample reward for his studies, and reduced again to indigence, which he had too keenly experienced formerly to render this a matter easily forgiven, Linnæus continued inflamed with rage against Rosen, who had stood in his way from his first entrance upon life. Nature could not heal his wounds, nor Friendship, for Artedi was away in England. No faithful sympathy could be found to soothe him, for mere fellowship in science does not serve us at the hardest pinch; nor could he disarm

[1] *Bältespännare.*
[2] In allusion to the thirteen 'nations' of the University.

the deserved blame of his elder friends. Although a good man and a pious, Linnæus was no saint. Heaven helped him, however.

While his desperate resolution lasted ferociously as ever, he awoke one night in agonising consternation from a hideous dream; he dreamt he had killed Rosen. He gave serious reflection to the horrid idea, and at length reason and religion calmed his violent passions. 'What an awful, wonderful thing a violent death is, even in a dumb beast!' says Kingsley anent the shooting of a horse. Oh, had he killed Rosen! The very thought was now agony to him. He felt he must go away and hide his face, his Cain-branded head, and (the Berserk fever over) sob out his thanksgiving that both of them had been saved.

From that hour he forgave Rosen; he even began to see that they two should be brothers in science. 'Shall God make us brother poets, as well as brother men, and we refuse to fraternise?' In grateful recollection of the impression made upon his conscience he wrote in after years a particular diary, called 'Nemesis Divina,' illustrating the words 'Vengeance is mine, I will repay, saith the Lord.' This is a small octavo pamphlet, written in Latin.[1] It contains meditations on texts of Scripture, Seneca, &c., and the self-searching of a penitent soul. That this penitence was not a mere passing impression is shown by the date of the 'Nemesis' pamphlet, August 31, 1739. The motto of

[1] Price seventy-five öre.

the book is *Innocue vivito : Numen adest*, which at this time replaced his personal motto, *Tantus amor florum*. But Rosen himself could not be reconciled: perhaps had they had an evening's chat or 'a morning walk together on their heathery moors, it would have brought their hearts miles nearer to each other, and their heads too.'

And so Linnæus and he still lived apart, both inspired with the same lofty aims, each following his star —the polar star—' *Se tu segui la tua stella*,'[1] success is sure—the essential is to see the star. They were to meet again. Carr tells us that Rosen, towards the close of his life, was glad of the medical aid of Linnæus: and the great botanist acknowledges frankly that he owed his life to the skill of Rosen. But much was to happen between this and then.

Linnæus does not in the diary mention having drawn his sword upon Rosen, though in the 'Nemesis Divina' he speaks of it befittingly. But he goes on in the diary to tell us, 'By this edict Linnæus was deprived of his only means of subsistence, and Rosen made up his mind to believe that he had now totally ruined him; but the following week there came a letter from Baron Reuterholm, with a bill of exchange enclosed, and a request that Linnæus would set out on his travels in Dalecarlia.'

No sooner was the door shut than the window was opened. Linnæus flew out of it like a bird. He was

[1] Dante.

not of those who lose an opportunity. The man who habitually loses his opportunities is sure to fail.

Here was the world opened to him again, the very best part of it, and life according to his liking. 'Because, though the whole earth is given to the children of men, none but we jolly fishers get the plums and raisins of it, the rivers which run among the hills, and the lakes which sit a-top thereof.'[1] For jolly fishers read naturalists.

[1] Kingsley.

CHAPTER IX.

ITER DALECARLIUM [1]—THE FAIR FLOWER OF FALUN.

> Up a thousand feet, Tom,
>   Round the lion's head,
> Find soft stones to leeward
>   And make up our bed.
> Eat our bread and bacon,
>   Smoke the pipe of peace,
> And, ere we be drowsy,
>   Give our boots a grease.
> Homer's heroes did so,
>   Why not such as we?
> What are sheets and servants?—
>   Superfluity.—KINGSLEY.

IT took some time to organise the expedition, which was no single-handed affair like the Lapland journey. This was a caravan of naturalists, to be furnished with due scientific and all other requirements—for the two young barons, Reuterholm's sons, could not be expected to travel without a certain amount of luxury. But to take servants and equipage were to imperil the objects of the journey; and Linnæus's fascinating tongue soon won over barons and all to trust to chance for their creature

---

[1] The adjective Dalecarlian is better Latinised Dalecarlicus, -a, -um; but as Linnæus wrote of his journey as Iter Dalecarlium, I have adopted his title. In Flora Dalecarlica the same. Linnæus wrote it Flora Dalecarlia. Jackson also follows Linnæus in this.

comforts, and carry only tools and necessaries. 'Knowledge is easily borne about,' said Näsman, the proverbial philosopher of the party.

For organisation Linnæus had a masterly talent, applying it to the arts of peace. He classified men as he classified other natural history specimens—by their qualities and capacities. He set about his work in a workmanlike manner.

He was accompanied by seven young naturalists, whom he selected from a crowd of volunteers. He enacted the laws and regulations of their course, and he appointed to every member his functions, to each one his manual and scientific as well as his administrative work. To each one a distinct department was assigned, and a report was given in at the end of every day's journey according to written rules which had been prepared before starting, for the due observance of which every member made himself answerable.

No writers of Linnæus's life have given us his travels; written in Latin or in Swedish, they have never been translated, except a few of them into German, and the Lapland journey into English. The unpublished journal of this Dalecarlian tour still exists, locked up, lying asleep, hitherto, in the Swedish tongue.

Honest old Stoever never professed to work upon the Swedish records, but he toiled diligently through the Latin (excepting what he could not get at in Sir J. Smith's collection), and with true German painstakingness—which seems a right German sort of word

—he has gathered together everything that was to be found in his time, and rolled it out in fine, respectable, nay more, genteel language, grandiose and flowery in the roundabout grammar of the period, when the ornate Louis Quinze style in rhetoric was thought the only thing fit for print. He gives this account of Linnæus's travelling companions and their functions, taken from the 'Hamburg News,' published some months later.

'*Näsman*, who had made himself known by a good dissertation on the Dalecarlian language, was to act as geographer; to give an accurate description of all the villages, mountains, lakes, rivers, roads, districts, &c.; to say morning and evening prayers; and to preach on Sundays.

'*Clewberg*, as naturalist, was to make observations on the four elements, such as on the quality of the water, on mineral springs, on sources, on the snow which never melts in the Alps in summer; on the height of the mountains, the weather, the fruitfulness or sterility of soil, &c., and to act as secretary.

'*Faldstedt* as metallist, besides collecting stones, minerals, earths, and all kinds of petrifactions (more generally loved in those days than now—Reuterholm especially affected them) &c.: as groom he was to saddle, water, and attend the horses.

'*Sohlberg*, an able student of physic, as botanist or herbalist, was to examine and preserve as well as possible all the trees, plants, herbs, grasses, and fungi.

He was also to precede the company as quartermaster, to procure them good lodgings, and to provide every necessary for their reception.

'*Emporelius* was zoologist, to describe and depict the quadrupeds, and all the animals living as well in the water as on land; also to shoot the game for the support of the company, likewise to fish.

'*Hedenblad* was to act as economist: to examine the dress of the Laplanders, their dwellings, way of preparing provisions, matrimonial and funeral rites, their knowledge of medicine, mode of living, diet, &c.; and to describe all this with pen and pencil' [Linnæus laid great stress on such graphic illustrations]. 'He was also to act as adjutant; to distribute the president's orders; to call the company together whenever it was required, especially in the evening, when an account was always given of the day's transactions; and to see that they went to bed and rose at the proper time.

'*Sandel*, an American, born at Pennsylvania, was steward and treasurer: he had the chief care of the fodder, cattle, wood-buying, and selling.'[1]

This worked well. In the space of a few weeks it

---

[1] C. Linnæus, Småland, Præses, publice in privatum.
Reinh. Näsman, Dalekarl., Geographus, Pastor.
Carl Clewberg, Helsingland, Physicus, Secretarius.
Ingel Faldstedt, Dalekarl., Mineralogus, Stallmeister.
Claud. Sohlberg, Dalekarl., Botanicus, Quartermeister.
Eric Emporelius, Dalekarl., Zoologus, Jirgmeister.
Petr. Hedenblad, Dalekarl., Domesticus, Adjutant.
Bemain Sandel, Americ., Œconomus, Körntmeister.
Iter Dal. July 3, 1734, Fahlun.

seemed as if they had been accustomed to this life whole years together.

The regulations are less remarkable than the fact that they carried them out to the end of the journey.

'The "Transactions" are printed (?) on forty-eight written sheets containing many important observations and discoveries. In the geographical part is a faithful description of the Dalelfven, the largest river of Dalecarlia, with all its arms and sources; also a geography of the Alpine mountains. . . . In mineralogy there exists a description of 120 different curious sorts of minerals and fossils, most of which are to be found in the district of Rättvik. In the botanical part is a list of all the plants growing in the whole province, called Flora Dalecarlia, with the synonyma and their economical and pharmaceutical virtues, written [1] by Baron Reuterholm.'[2]

Although Gieseke wrote as above, the 'Iter Dalecarlium' never really seems to have been printed under the superintendence of its authors. It was consulted in MS. by Linnæus's pupils, and the botanical remarks were inserted in his own printed works. The journal seems to have been used, artist fashion, as a quarry for materials. One particular fruit of this journey was a list of the pasture herbs of Sweden, published under the title of 'Pan Suecus.'[3]

[1] Collated and transcribed.
[2] Article on the *Iter Dal.* in the *Hamburg News*, written by Gieseke.
[3] In *Pan Suecus* are recorded over two thousand experiments to

'When the president discovered a village it was not necessary for all the company to ride thither, but the geographer alone was sent to enter it. If some particular stone or fossil was found on the way the metallist was directed to alight; at the sight of some curious plant or insect the botanist or zoologist did his duty. They took the respective objects with them, and prepared a description, to be inserted at night in the "Transactions," besides the name of the place where they had been found. At night they all met together; the president then dictated to the secretary[1] the memoranda collected by each companion in a regular turn, from the geographer to the steward; and if he happened to forget any remark, the companion to whose office that part of the science belonged refreshed his memory.'

This division of labour was found easily practicable by the party. All did not precipitate themselves on a

---

ascertain what plants are eaten and what are rejected by horned cattle, goats, sheep, horses, and hogs. Linnæus conceived the design of instituting these experiments when he was travelling in Dalarne. Horned cattle ate of the plants offered to them 276 species, and rejected 218; goats, of 449 kinds, refused 126; sheep, of 387 kinds, refused 141; horses, of 262, refused 212; swine ate of those offered 72 kinds, and refused 171. Three-fourths of these (Swedish) plants are the same as ours in England. The utility of such experiments is evident, as they lay the foundation of further improvements in the economy of cattle. Our hay might be much improved; for although cattle will eat in hay those herbs which they reject while green and growing, yet it does not follow that all in their dried state are equally nutritive and wholesome.

[1] Gieseke. This may have been the original plan, but in the journal each man's observations are recorded in his own handwriting.

plant, nor was there a struggle to obtain the best or only quarters for the night, as I have seen a party of thirty 'géologues' in France simultaneously claiming the only rooms to be had in small French townlets, and clearing the neighbourhood of its *café au lait*; jolly, very, but verily unsystematic. This was likewise a 'jolly-gizing' trip, as the livery-stable keepers called Professor Sedgwick's adventurous rides with his train of pupils. Linnæus set the fashion of such excursions, picnic jaunts, in which he and his pupils enjoyed nature, capping verses or quotations as they rode, keeping their eyes, minds, and hearts open. Youth and health are never at a loss for laughter. Young Faldstedt, the athlete of the party, who groomed their horses, acting, as he said, as Master of the Horse, was a playful, impudent, careless, jovial, capital fellow, always keeping their energies up to the mark; Näsman was graver, as befitted his pastoral character, but always ready with a Swedish proverb, which Sandel, the Pennsylvanian, capped by a world-wide or smart American saying.[1] Linnæus ruled his little troop well, being by nature superior to the rest of them, who were, however, of a high average. These young, pliant, susceptible natures had their lives coloured by his friendship, and their minds moulded to his by contact with his clear thought and elevated feeling.

The air, too, is so fine in these inland parts of Sweden

[1] I have thrown the dry materials of the reports somewhat into narrative, and even ventured to introduce an occasional dialogue, founded always on the substance of the journal, unless it is for the purpose of introducing a Swedish proverb.

as to brace and stimulate the faculties, and produce that state of wholesome intoxication which affords its pleasures without its penalties. Guthrey says that 'the inhabitants live so long as to find life tedious, and therefore go to other climes of less salubrity.' Never mind, they would find less inconveniently fine air in some of the mining districts where they were going; mayhap they would even find diseases. Their object was to go to Röraås, high up in the dreary mountain range dividing Sweden from Norway, where they might chance to die of cold and privation, aided by mining dangers, arsenical fumes, and others. Though the copper exhalations are so deleterious to vegetation, they seem to offer some hygienic advantages to mankind. Falun, like Swansea, has always been exempt from cholera and other pestilence, and this immunity is attributed to the smelting of the copper ores. This is considered an excellent disinfectant.

Sohlberg, a former pupil of Carl's at Upsala, one of the Dalecarlians of the party, led the way, because he knew it, and because his office was to quarter the troop comfortably on their arrival at the end of the day's journey. The rest rode more at ease, making busy notes for the journal, whose early pages are naturally the fullest and neatest. There are different handwritings in each leaf, and a broad margin on the outside of each page, which Linnæus fills with his comments, which show how carefully he read every day's report.

The papers each day begin with the geographer's

report, then the botanist's (Sohlberg, whose day's public work is over early). The American 'economist' makes the most drawings, even more than Hedenblad, whose especial function it is; he makes sketches of tools, appliances, and anything that hits his fancy as a 'notion.' The first day's report is the neatest of all. It is a general experience; one writes most, the fullest diary, on the day of leaving home; one's blood is up and one is not yet tired; all is hope and expectation. Näsman, the geographer, made a large careful map on time-browned paper, marking in water-colour the streams where they rise from lakes looking like leaves at the end of branches. The churches are all marked as far as Idre. The map itself goes no farther up.

They left Falun on July 3, 1734, the seven Dalecarlians (including the two young Reuterholms) as eagerly on the watch as the others for natural objects as yet unobserved by them. They were eager to distinguish themselves in the diary.[1] The high road to Leksand[2] just after leaving Falun, covered with masses of mineral refuse, and bare of vegetation, presents much the aspect of the bleaker Cornish mining districts, with the range of the Stora Kopparberg forming a dreary barrier between it and the pleasanter parts of the world. There is a luxuriant vegetation outside Falun beyond the range of the fumes from the smelting-works. The mines of these 'big copper hills' have been

[1] The Reuterholms never write in the journal.
[2] Spelt Lixan in the MS.

worked over six hundred years. For the more complete exploration of the province they occasionally divided themselves into two bands. Some of them now took the northern road by way of Biursås on the Rögsjö. The rest travelled by way of Junsta and the lakelet of Innsjö. The peculiar and primitive population of these valleys are among the handsomest of the Swedish race; they still retain their national dress, and are proudly independent in their faithful adherence to their ancient customs. This was less remarkable to Linnæus than it is to us, as at that time the Swedes in general resembled in costume and manners the Dalecarlians of the present day. A long boat loaded with about seventy of the Dalfolk, coming across the Innsjö to Brednäs with a wedding-party, seemed a signal for a halt, and our travellers joined the holiday-makers in the spirited and delightful Dalecarlian dances.

The four most interesting parishes of Dalecarlia are Leksand at the southern and Mora at the northern end of the Siljan Lake; Rättvik, at the end of the large bight, extending north-easterly; and Orsa, on the lake of that name, which is, in fact, a part of the Siljan before its volume is increased by the tribute of the Öst-Dal River. The population of these valleys is about 170,000. Besides their farm-work they make extensively (in their own houses) basket-work, clocks, watches and tools; likewise bells, furniture, grindstones, &c.; in all of which Sandel the *œconomus* was vastly interested—possibly because some of the appleblossomcomplexioned female workers were so pretty! The diary does not say so, however. The frequent farms and villages of red and white painted houses set in the sylvan scenery of wood and dale, and the numerous boats on the silvery lake, make the soft and beautiful scenery animated and still further interesting. The costumes of the people are charming. The red bodices of the women rowing the long boats, reflected in the glassy Siljan Lake, are delightful in colour. Linnæus's whole party were hard at work observing, collecting, and taking notes, alternately riding and boating. The country, being nearly one-quarter of it covered with water, seemed to leave to Emporelius, the fisherman and zoologist, the lion's share of the work of observation. Still, the remaining fourth, intersected with brooks and feeders to the lake was 'just the sort of country to learn something in.' The yellow anemones were over, and all in seed, but the delicate perfume of the *Linnæa borealis* filled the air. The youths saluted this graceful plant [1] by the name it was thenceforward to bear, though the name was only printed on their hearts as yet. Wild strawberries and raspberries grew abundantly for their refreshing.

> So they passed by in their joy, like a dream, on the murmuring ripple.

Some of the party rode round the north bank of the lake, passing through Rättvik, leaving Clewberg, Empo-

---

[1] *Campanula serpyllifolia*.

relius the fisherman, and Sohlberg the quarter-master, who could not obtain horses, to come on by boat. The gay costume of the Rättvik women is as pretty as any—white sleeves, blueish skirt with green border and dark waistband, a woollen apron with transverse bands of white, green, yellow, and blue, embroidered leggings, the cap black with red trimmings, or sometimes of linen, with two balls falling on the back. One frequently sees this costume in Stockholm now, as well as other varieties of the Dalecarlian national dress.

The first four stages of their journey lay through a hilly well-wooded country, of which Wickarby, lying beyond the handsome church of Rättvik, is the most delightful district in point of scenery. The views from the hill-range called Bergsängsbackarna are extensive, and present with great attraction the varied charms of wood and water, the scenery becoming wilder as one penetrates the valley to Ofvanmyre and Boda, where there is a lofty waterfall (200 feet). The party formed a junction at the copper-roofed church of Mora, where the scenery becomes tame, and travelling tiresome on the sandy road. Linnæus with most of the party again turned off at Mora Noret, where the Orsa Lake in a broad stream empties itself into the Siljan; then took the eastward road to Orsa, forming a junction with the others where the roads join before Garberg, keeping the East Dal River to the left. He did not wish the young Reuterholms to be fatigued with the longer exploring rides, so he left the boys in Näsman's and Sohlberg's care to travel by the shorter road.

The active well-grown handsome farmers of these parts interested Sandel, the American, particularly; they can turn their hands to anything, having generally a trade—such as blacksmith, tailor, or what not—in addition to their farming. Hedenblad made notes of their costumes. The men wear short coats of white homespun cloth, generally two of these at a time, the under one being sleeveless; they have white leather breeches and blue stockings. The women wear bodices of scarlet wool, showing their long white linen sleeves, a dark blue cloth shirt, and a yellow apron with a black border, white stockings, and shoes with a peg-like heel in the centre. These people are tall and well-grown, and altogether a fine race, which makes it the more remarkable that they, the men in particular, are said to seldom outlive thirty years of age. This is partially explained in several ways. In the notes on domestic medicine in the journal pleurisy is mentioned as a distemper of an epidemical nature in that country; it is alleged that it arises from the excess which the inhabitants commit by gorging themselves with a kind of porridge made of flour. Hedenblad alludes, besides, to their mastication of a certain kind of rosin, and describes their burying in the earth a certain species of rotten fish, called *Lunsfisk*, which they dig out again to prepare it for their food. They also drink the strong Norwegian wine made of berries. Linnæus is of opinion, with regard to the early deaths at Orsa, that the mortality is due to hectic fever arising from the pernicious exhalations of the mines. He thought he found a still better explanation of the disease when riding through the narrow streets of Orsa Kyrkoby, where sixty or seventy farms lie very close together, and the close unwholesome stabling attracts swarms of flies. He was inclined to think here lay the root of the evil, the extreme closeness of the dwellings and the sudden exposure to cold air precipitating the disorder. Lunden, a hamlet by the shores of Orsa, is a wholesomer, pleasanter place than Orsa Kyrkoby (church town), where the dazzlingly white and clean parish church is a fine example set to the housholders. The people here are poorer than at Leksand. It is good and pleasant for them to get away for the summer months to the *fäbodar* (*saeters*), where the marshy pasture-lands afford rich feeding for the cattle, and the ponds, fringed with horsetail,[1] are filled with clean water, while the air is filled with the fragrance of the pines. One hears the girls calling the cattle with a horn, or sees them knitting stockings, surrounded by the herd, which they lead readily by means of a pocket of salt at their waists.

For their own part the exploring squadron preferred to sleep in a barn on new hay, or on bags filled with the dry Dalarne moss. 'These mosses of the Dalarne woods are all in some way remarkable. *Fontinalis antipyretica*, the longest of the tribe, is much used by the peasantry as a remedy in their chest complaints, also as a preservative against fire. The farmers place it between the stones

[1] *Equisetum*.

and wooden walls of their houses; like asbestos, it will neither light nor retain light.' The young men's evenings spent quietly in the barns were as pleasant as those occupied in dancing. 'With them talk was what it ought to be—an exchange of information, thought, and argument.' The party exchanged their news, having again formed a junction: the eager enjoyment of novelty was still felt by all. While waiting eagerly for their turn to write their notes in the journal they came outside and left the place silent and undisturbed to the immediate writer, while they relished that greatest enjoyment in life, 'intercommunion of equal minds and sympathetic hearts.' They had passed a studious winter, and 'headwork demands physical relaxation.'[1] How to develop the physical powers sufficiently for making the very best of life's work 'without engendering brutality and coarseness,' is the puzzle of this age. Linnæus, the Kingsley of his time, seems early to have solved it. 'Plant-hunting was to him what sports are to other persons.'[2] While himself 'in ecstasies of observant study,' 'this robust genius, born to grapple with the whole army of nature and to marshal it,' was yet able to train the young men who formed his school—an anticipation of his future court—to guide rather than to rule them by the life, the pleasure, the intensity of interest he infused into every object, and by his sympathy with the opening of this revelation to them, and with their new enjoyment of the 'magnificent smile of mother nature, most genial but

---

[1] Kingsley.   [2] Bain on Mill.

most silent.' In view of his after career, the experience that Linnæus now gained in training the attention and winning the affection of young men was the most valuable result of the Dalecarlian tour. When the young Reuterholms complained of the great length of the Swedish summer days, Linnæus was prompt to show them how much their country was to be envied for this very thing: how two great batches of work could be worked off in each year in Sweden, such as could not be steadily performed where time was more broken into. They had the long summer days for discovery, the long winter nights for classification. At least this is what he seems to mean when he speaks of the dark winter days as such an advantage of the Swedish climate. Linnæus had pre-eminently the faculty of being wide awake—not the highest endowment by any means, but the most useful for his purpose. Linnæus never lost himself in dreams: he was always more in the body than in the spirit. He never thought, like the poets, of seeing the invisible things, but he kept a keen look out for things visible but as yet undiscovered, while his ears were awake to Clewberg's, and more especially to Emporelius's and Sohlberg's, tales of discovery, related with all the fire of youth. 'Every fly that lit upon the boat side, every bit of weed that we fished up, every note of wood-bird, was suggestive of some pretty bit of information on the habits, growth, and breeding of the thousand unnoticed forms of life around.'[1]

[1] Capt. W. Congreve.

They changed horses at the pretty hamlet of Garberg, 'pleasant with red houses, verdant fields, and groves, and a stream of clear water.'[1] The next hamlet of importance is Elfdal, near which are now celebrated porphyry works.[2] Faldstedt made a special report upon them, while Hedenblad's attention was drawn to the signs of Lapp ancestry among the people, who are short and ugly, at only fifteen miles' distance from the finely-peopled parishes of Orsa and Mora. The scenery round Elfdal is very picturesque. The numerous falls and cataracts formed by the Dal river add much to the beauty of the landscape. Henceforward traces of human industry become less frequent; mountain, ravine, cataract, and pine-forest follow each other in endless succession. The shooting in these forests is highly spoken of—bear and elk, capercailzie and hazel-hen, and most kinds of game. Emporelius's gun was constantly at work for specimens; he fed the party well besides. Ten miles farther up is Åsen, with a chapel, where the pastor received them for the night.

They crossed the East Dal at a ferry near Åsen, keeping the river on the right hand, and after a long ride they rested at a large hay-house. From Åsen to Sürna[3] it is six Swedish miles, with no hamlet or post-station between these places. This long march from Elfdalen to

[1] Du Chaillu highly praises this landscape.
[2] Linnæus's monument in Upsala Cathedral is made of the red porphyry of Elfdal.
[3] Spelt Serna in the MS. diary. The spelling of proper names is unsettled throughout.

Särna is cheered by the fact of the road passing through the wildest and most magnificent scenery of the whole pass. The cattle-bells tinkled and the herd-maidens played delightfully on the horn. While Emporelius was fishing in the Elf,[1] and having the eatable specimens he had shot prepared for their supper, Sandel, the American, and the young Reuterholms amused themselves with improvising a concert and improving their own method of calling the party together by the sound of the horn. This instrument, however badly played, never failed of its effect about meal times; at bed and getting-up times it was less successful. The boys said Hedenblad had feeble wind—he must practise longer to strengthen it; and they wondered what profession would admit of their lying in bed the longest. They had to be up early on the morrow, for they had loitered a good deal on their way. Not until July 14 did they arrive at Särna. They set their three-quarters of a dozen watches and went to bed. Linnæus alone had no watch; he knew the time of day or night by the birds and flowers. 'Hasten them up, it is late,' said the leader on the morrow: 'it is five o'clock; the yellow mountain poppy has just unfolded. The blue-throated robin woke the day three hours ago.'

Särna is prettily situated on the river, which here widens into a lake. The pastor housed part of them in the parish church, the inn—even then a good one—being full on account of a fair that was about to open. Särna

---

[1] Or Dal, or Dal-elf, river of the dale.

Lake was on their right, but they crossed the river again above Sürna, keeping the Idre Lake on the left and the Stadjan Mount on their right. It is three Swedish miles hence to Idre—a poor hamlet, with scattered farms. The present carriage-road ends here. It was merely a bridle-path in Linnæus's time. The forest path leading to Norway was then the roughest of tracks. They could not make much headway on horseback through the forest; they found it easier to dismount and lead the animals.

'Better so for my trade,' said Claes Sohlberg the botanist, holding up a prize specimen of *Andromeda florecœruleo*. This was on July 16.

'He who is afraid of leaves must not come into a wood,' said Näsman the proverbial philosopher.

'I expected to find so many more birds as we came to the deeper forests,' said Emporelius; 'we never hear a song here.'

'That is because autumn is coming on,' said several of them. 'Sweden is always silent in autumn: one hears only whisperings from the wood.'

'It is a general mistake,' said their leader, 'to expect to find song-birds in the deep forests: they cannot find their food there; while their enemies the larger birds— hawks and owls, that feed on flesh—are more abundant.[1] The song-bird is the friend of man, and loves his neighbourhood, which includes his corn-fields and fruit-gardens.'

[1] Lowell also notices this.

'A thousand probabilities cannot make one truth,' observed the sententious Näsman.

Pine woods are generally said to be silent, nevertheless Sweden in spring is highly musical, indeed quite orchestral, with birds.

They came out to an open space where were willow-grouse in abundance. Close time was not then and there thought of; Emporelius had levelled his gun for a shot, when he was stopped by the sight of an unfamiliar species of kite poised motionless above. He wavered and missed everything.

'All covet, all lose,' quoth Näsman.

'I wonder we never see the red grouse of England and Scotland here,' said Sandel, who had stayed in England on his way from America. 'The feeding would be much the same.'

'It may be the same bird as our grouse, changed somewhat by the conditions of our climate,' said Linnæus.[1]

The red grouse of the Scotch and Welsh hills are the only large and conspicuous creatures entirely confined to the British Isles. All our other animals have come to us from somewhere else, but the red grouse is found nowhere else than in Britain. The only bird at all closely resembling it is the willow-grouse of the Scandinavian peninsula, which changes its plumage

---

[1] Linnæus classed several species of birds in the genus *Tetrao*. Dr. Bree considers that the affinities of the willow-grouse 'are more with the ptarmigan than the red grouse, but it is distinct from both.'

annually with the approach of white winter. It is probable our bird closely represents the common ancestor of both species, which must have come northward into the unoccupied hills of Scotland and Norway when the vast glaciers of the great ice age began to melt off the face of sub-arctic Europe. In proportion as each northern grouse grew lighter and lighter during the winter season would its chances of escape in the struggle for existence grow ever greater. The darker-coloured individuals would thus at last entirely disappear, being one by one weeded out and annihilated, while the white alone were left to form the parent stock for future generations.[1]

The dark fringe of the pine-forest rested on the white cloud-masses above the snow-topped hills of the lofty Slerol Stadet.

> How sharp the silver spear-heads charge
> When Alp meets heaven in snow !

The mountain precipices at times overhung their narrow pathway as the discoverers followed each other in line, some looking professionally upwards to the hills above, some downwards to where the clouds floated in dense masses below their path on the left hand, others curiously investigating the walls of rock on the right, which were as if fresco-painted by the variously coloured lichens that climbed, holding tight by their tiny teeth, while other

---

[1] Condensed from *Times*, September 5, 1885--'Grouse of Great Britain and Ireland.'

mosses seated on higher crags expanded themselves and wrote their history in capitals on those giant walls. 'Wit in a poor man's head and moss in a mountain avail nothing,' said the poor curate Näsman. Linnæus contested this.

'The stag's-horn club-moss ceased to straggle across the turf and the tufted alpine club-moss takes its place: for they were now in a new world—a region whose climate is eternally influenced by some fresh law (after which Clewberg vainly guesses, with a sigh at his own ignorance) which renders life impossible to one species possible to another.'[1] The scenery and its solitude would have been oppressively magnificent had it not been for the stimulating quality of the air, which raised their energies to meet the demands made upon them. Their minds were set in full tension of receptivity, a new intellectual world seemed to unfold itself before them, wooing them to its conquest. It was like awakening out of a night of ignorance; and they were at the end of the day's ride, and Faldstedt, who had disburdened his pockets of their mineral collection, was already leading off the horses to shelter before they had exhausted their questions to Nature and their leader. Without actually expanding in verse—an instrument few of them had cared to practise—the grandeur of the landscape, and the excitement of being the first to examine it, set these ten young fellows all glowing into poetry at once in the joy of discovered relationship with infinity.

[1] *Glaucus.*

> Are not the mountains, waves, and skies a part
> Of me and of my soul, as I of them?
> Is not the love of these deep in my heart
> With a pure passion?

'We sprang from earth, the first and last home of our bodies, our souls (the habitation of the spirit) evaporate to the skies; our whole nutriment is drawn from earth.' And these young men were now first winning their mental nutriment likewise from the lap of Nature.

The sun was hidden: the pine-tops against the blue and white clouded sky became a view in black and white as the light went out. The sun burst forth again: the picture became a painting. The lyrist Clewberg, the only actual versifier of the party, was putting his feelings—gifts from the greater world, he called them—into tender sonorous words, when the horn sounded and Linnæus called them to a demonstration on the material aspect of the world immediately about them. Perhaps this was even more poetical than Clewberg's unwritten poem, which began as an invocation to divers forest nymphs. 'Pooh! nymphs? We want no nymphs,' said Linnæus; 'our mother Nature's beauty and beneficence are enough for us. Let us keep fast hold of her apron-string.' Modesty and deference—'virtues of sacred obligation,' inculcated by Näsman, who, insisting upon good manners, always said, 'Do on the hill as ye would in the hall'—prevented the younger men laughing at the idea of Linnæus never intending to fall in love, nor having yet done so.

'He'll find out by-and-by that it is the nature of man to fall in love, and then he will have that sort of nature to study,' whispered the elder of the two Reuterholm boys, who fancied himself grown up, to Hedenblad, who had prepared some careful notes on the matrimonial rites of the Laplanders. Hedenblad again wound his horn for the lecture.

Linnæus was standing on a rock making observations on the weather from the movements of the birds and beasts. There had been a long drought, and now evidently something was taking place betokening a change. The creatures were restless yet not shy. The golden plover piped loud and long. The swallows were skimming the Idre Lake quite close to its glassy surface. 'We need carry no barometer,' said the leader; 'there is always one ready to our hands on these fells. The tools are always here; we have to learn to use them. However the clergy may construe it that the kingdom of heaven is within us, the kingdom of earth certainly is so, and education's business is to open up this kingdom to us. There, Emporelius's shot has broken up my barometer.' The sharp report sent with a start thousands of winged things all into the air at once. The sportsman took up bleeding and still palpitating one poor little willow-grouse in its summer dress, but with its wings, breast, and legs still unusually white. 'The rest are more frightened than hurt,' said Näsman sarcastically.

It seemed as if the shot had brought down the rain

with the bird, for a heavy shower, 'raining grass and gold,' came down, diluting Clewberg's ink and driving the young men to shelter. The hut, though dark, was anything but weathertight: the chinks and crannies were 'all squirts and whistles.'

'A child may have too much of his mother's blessing, eh, Näsman,' said Clewberg, laughing as the candle hissed and went out, and he could not see to put his ink under cover. Two bats—one the *Vesp. borealis*, the other a long-eared bat—whirled inside the hut. They were caught that night.

'Change of weather finds discourse for——'

'Englishmen,' finished the youth, merrily snatching the rude Gothic proverb, which says 'fools,' out of Näsman's mouth, who was peevish at having his province of proverbial philosophy invaded.

'After clouds comes clear weather,' said Näsman, recovering his temper, and the two left the hut together.

They saw a strange sight. The short shower, which had been local, and focussed in their immediate neighbourhood, was over, but in the distance the vast mountain wall was partially illuminated by jets of flame, spouting upwards, or running about waywardly on the ground, and now and then shooting or darting forward like fireworks. 'It is the wildfire,'—'It is the *ignis fatuus* playing over those deep marshes,' were the utterances, awed or philosophic, as they watched these phantom fires. The thickening darkness intensified the effect, which was amazing, and at times horrible. Every now and then an explosion would be heard from a newly lighted flame, and each explosion brought down a burst of rain. They listened for thunder, but only cracking reports were heard, no long thunder-rolls. All at once the great mountain mass of the Slerol Stadet stood out black and terrific against a background of blue flame, and meteoric stones shot hissing into the pools of the marshes. 'Is it the Aurora Borealis?' whispered one of the young Reuterholms—they had all gathered round Linnæus for safety. He could not tell, he had never seen anything like it before; it was some awful convulsion of nature, but it was not that. They must trust God's mercy for protection.

One great clap was heard, thunder at last—real thunder this time—following a blaze of lightning so vivid as almost to obscure the deluge of blue flame now rolling round the base of the mountain, and down came the rain in torrents, driving them all to the shelter of the hut. The drought had broken up in a fearful storm.

But the storm could not explain to Linnæus the meaning of those wayward fires, whose terrors were seemingly soon forgotten by the rest. He, who alone had appeared calm among them, could not conceal from himself that here were real terrors, and something more than met the careless eye. The storm was over. He wandered out into the valley all alone, carrying a lantern which cast a fitful ray over the warm, vapoury silent wilderness. He was the only moving object. There were neither stars nor moon, only the moist steamy dusk of a Lapland summer night when the air is still but not clear. With cautious steps he moved downward to the marshes: he could not return to the hut to rest, for he felt on the eve of a discovery. This is what he says of what he saw as morning light helped him on his search—he says it later on in life, to the crowd of listening students at his first lecture [1] as a professor: 'You will scarcely believe me when I tell you that there are whole mountains full of petroleum in Dalecarlia. Yet doubt not. This thing, hitherto unheard of, unseen, I myself saw with these eyes, and was surprised.' The terrors of the frown of nature were over; only the aspect of her bounty remained. 'God is always opening His hand,' said Näsman, who had also come out to view the scene. This, then, was the explanation of the terrific explosions of last night. The great heat had kindled the fire which fed upon the parched surface of the peaty ground. The thunderstorm had gathered force in time to prevent a vast general conflagration.

The account of these petroleum discoveries is scarcely to be read in the pale and panic-stricken writings of July 17; on the 20th they blackened the ink down, or procured some more from a priest; and Sohlberg's nerves were by this calm enough to classify his plants and catalogue the vegetable products of these parts—including *Campanula serpyllifolia*, the *Linnæa borealis* that graces these wilds—as well as to note the great size of the trees forming the log-huts, those at least whose engraven runic

---

[1] *Oratio de Peregrinationum intra Patriam Necessitate.*

characters showed them to be very old. Trees seldom grow to such dimensions now in Scandinavia.

The river is divided into three branches, two of them flowing from lakelets or mere ponds; the third branch, which the travellers followed, flows from the Fæmund Lake. Logs placed side by side across the bogs enabled their horses to pass the swampy tracts. They halted at a small farm on the shore of the Elgsjö (Elk Lake), which they reached by a ferry across the stream, which is here unusually deep. 'Be ye the last to go over a deep river,' said the proverbial philosopher, more cautiously following Clewberg, who had the ill-luck to be soused; and there he stood, 'all shivering and shaking, and the water a-squish-squashing in his shoes, and his trousers all sticking slimsey-like to his legs,' as with fumbling feet he crawled and splashed out, a magpie all the while eyeing him keenly, with the instinctive humour of animals. The magpie was a find. It was a Dalecarlian magpie never described before, 'whose feet are not armed, like those of other magpies, with four claws, but have only three—two before and one behind, which is rather stronger than those in front.'[1] 'He's laughing at you, Clewberg,' said Sandel; 'he's asking you about the quality of the water.' Emporelius's gun had already made an end of the poor magpie who had presumed to quiz a philosopher, and that very night he was prepared as a specimen. But the sportsman's triumph was of no long duration—his ammunition had run out.

[1] Lowell.

He had left a large depôt at Särna to take up on their return journey, and could only depend upon a chance supply until they should reach Röraås. Hungry as they were, they must live on fish till then. The higher up the mountains they came the hungrier they grew, and the keener the relish they had for animal food.

'It is too provoking, just as we come upon the line of the reindeer, that we can't get it now while it is so nice and fat,' grumbled the hungry ones.

'Dearths foreseen come not,' said Näsman reproachfully to Emporelius.

'We have a proverb in Småland too,' said Linnæus; 'that "one is sure of a supper if there's plenty in the knapsack." Now let us all go out and fish.'

'The river is swarming. They are flopping and smacking about in all directions; but, oh, dear! why did Heaven make midges?' said Reuterholm major, complaining (like our Kingsley) of Nature's inexhaustible opulence.

'For our collections, to be sure,' said Emporelius, the zoologist.

'Näsman says nothing,' remarked Clewberg. 'He is thinking of "After meat, mustard."'

'A close mouth catches no flies,' mumbles Näsman, drawing on his gauze cap.'

Emporelius was no busier than the rest of the party in completing his insect collection. Before supper each one of them had slain a hecatomb of pertinacious flies.

The insect collection made in this expedition is very

fine. In Linnæus's tours in Lapland and Dalecarlia he gathered 1,000 species of insects, which are neatly pasted on paper; among them are sixty-five different specimens of flies, including the large fly the *œstrus*, which makes such havoc among the reindeer in Lapland.

The Linnæan Society possesses Linnæus's insect collections. They are in damaged condition, as Sir J. Smith was less of an entomologist than a botanist, so that there was no care given to arranging the specimens.[1]

'Nature seems exhaustless in her invention of new insects hostile to vegetation.'[2] Sohlberg found none but imperfect specimens of leaves, which disturbed him, as, to his surprise, he found the leaves of broad-leafed plants, such as plane, and maple, and some poplars, become much larger as they journeyed northward, while the trees themselves were unusually small.[3] Linnæus accounted for this phenomenon, which he had himself observed in the north of Norway, by the long duration of the daily light in summer.

They passed Fiellen, continuing their way through the forest, the refreshed mountains being now 'silver-veined with rills.' The clouds on the highest mountain, Slerol Stadet, which had first appeared below, now approached the travellers, writes Clewberg in the journal. They skirted the reed-fringed Storbosjö and came to

[1] There are reckoned to be 112 species of butterflies in Norway, Sweden, Lapland, and Finland.
[2] Lowell.
[3] Much of this country belongs to the willow and birch region.

Norvig, at the head of the Fæmund, which is a swollen stream, called here a lake. Near this point there is a pass to the left, 'where a rough path leads to the high road to Röraås from Hudviksvall on the Baltic.' Here they crossed the Norwegian frontier.

The Fæmundsjö is 2,150 feet above the sea and 35 miles long. This beautiful sheet of water has peculiar scenery not possessed by any other Norwegian or Swedish lake; its shores are not abrupt, and in many places they are thinly clad with fir, and birch, and fine reindeer moss. The clear water swarms with fish, and wild reindeer browse upon its shores.[1] Its outlet, the Klara (clear) river, flows into Lake Venern. The exploring party made their way through the defiles of the mountains and looked upon the sister kingdoms.

On July 30 the geographer, Näsman, gives a marginal sketch of the rivers and lakes and Dalecarlia generally, as he 'pervestigated it' (1734, Dalekarliam occidentalem et orientalem pervestigavit). On July 31 Emporelius, the 'zoologus,' makes a sketch of a reindeer—as if they first saw reindeer that day. The youths all shouted as they saw the comical portrait in the book. The intense stillness was broken into echoes by their shouts. Emporelius modestly admitted he was no great artist as the others were. He drew the lichens and the reindeer-moss—these much more carefully and delicately in the diluted ink. They now stood on that highest point where there are no more hills, but one looks below into a

[1] Du Chaillu.

vast hollow fringed with blue, the border of which basin sometimes rises into a purple battlemented wall. They soon began the descent. Here man stepped into the scenery and set his huts among the great gables of the mountains. How dwarfed and ugly are all human works in a place where even the Parthenon would look like a packing-case! The huts at Röraås are infinitely mean and ugly. It is a poverty-stricken Falun tossed 2,000 feet into the air.

The interiors are more comfortable to view, and here men resume their natural proportions. Du Chaillu describes them as 'men in knee-breeches, white woollen stockings, double-breasted waistcoat with shiny brass buttons, and red Phrygian caps.' The hut that housed him holds 'old porcelain dishes and cups—heirlooms; a lantern hung from a beam in the ceiling; an old clock near the bed.' Hedenblad and Sandel take no notice of these every-day details. The copper mines at Röraås have been worked since 1644. The town is 2,000 feet above the sea level; the Storvarts mines 2,800 feet.[1] The Hitter River flows through the centre of the town, the two parts of which are connected by wooden Norwegian bridges. The large church was not built in our travellers' time; but the woods were then more abundant and the climate milder. The wolf and the glutton, the reindeer's greatest enemies, were common in these woods. At present the country round Röraås is remarkably bare and bleak.

The young naturalists found the Norwegians dirty

[1] H. Marryatt.

and grasping. 'Bread with eyes and cheese without eyes,' quoth Näsman. He recommended his companions to 'be ready with the hat, but slow with the purse.' 'A small sum,' said he, 'will pay a short reckoning.' He tried it in vain.

The sturdy Norwegian laughed at his civilities and held stoutly to his bargain. 'Take it or leave it, but I won't take an öre less.' Yet a minute afterwards he good-naturedly offered the whole hospitality of his house to Faldstedt, the metallist, limping beneath the weight of his specimens. The party remained four days at Röraås, examining the ores and mining-works, and after making an expedition of one Swedish mile to Grüfum, they began their homeward route. The birds, too, were all preparing a farewell to Röraås. The expedition, however, did not follow the swallows' flight due south, but, re-entering Sweden at Sverige, they followed the rough uneasy tracks of the West Dal, so difficult to find one's way in. They rejoiced when the evenings drawing in enabled them to travel by the polar star, called by the nomad Turcomans the Iron Peg, because it holds so firm. On August 4 the botanist Sohlberg speaks with some surprise of the large potatoes: an exception to the general rule here, where 'potatoes usually grow so fast that the tubers are small, all the strength going into the stem.'[1] On the 5th he rejoiced to find the stately plant (*Pedicularis*) called the *Sceptrum Carolinum*. Linnæus had already met with it in Lapland.

[1] Du Chaillu.

On August 7 the reports are all very short, hurried, and so cramped as to be almost illegible. It was a hard day that, as the 6th had likewise been. They came to the Särna Lake, where the East and West Dals rejoin and again diverge. The first church on the return journey is Transstrand. Here the road begins; a mere track it was then. Then comes Lÿma Kyrkia; Malungs and Eppleboda (Appelbo) are the next hamlets. The inhabitants of the West Dal are a quicker and more lively race than those of the East Dal—so much so that they seem to be of different origin. The travellers still keep up the journal with its observations, but it has lost the spirit of the earlier record. They were all tired, by day only eager to get on, at dusk only eager to turn in. The Lapland bunting with his single call-note failed to waken them early. Linnæus alone stood ever on the watch, like Columbus on the prow of his vessel, eager for fresh discoveries, patient, brave. But this was easy work to him after his toil in the Lapland Alps. As they get on lower and lower ground the days draw in perceptibly. 'The sun's rim dips, the stars rush out. At one stride comes the dark.'

They all wind their watches, implying by their action that they have well filled that day and may roost content. Their leader is watching the sun and moon, the mighty timekeepers. He is never contented with himself, as the world in general is never content with things outside. An early frost has blackened the potato patches and tinged the beautiful flowing hair of the birch trees with pale gold. The game-birds crowd into the sunny spots in the early morning to warm and dry their wings. The roads are soft and very bad. Next morning a sharper frost strung the boughs of the birch trees with icy pearls. Each leaf was outlined in white hoar-frost crystal; sharp splinters of frozen dew turned the pine-needles to rays of light, glittering like spar; while by the riverside the wet willow branches and the alders were hung with icicles, which shook and rung against each other like frost-fairy bells above the blackened river, which was covered in smooth places with ice quarter of an inch thick. The road, however, was good for travelling on.

> I [winter] make causeways, safely crost,
> Of mud, with just a pinch of frost.

Lowell's phrase applies to Sweden still—' Winter is the mender of the highways: every road in Europe was a quagmire during a good part of the year, unless it was bottomed on some remains of Roman engineering.'

With the changing of the sun's crimson into yellow and then into intense white, vanished also the fairy-like appearance of the crystallised groves of birches; the air first grew steamy and then hot. The ground was empurpled with the heather.

The returning party kept the way down the noble valley watered by the West Dal River to its junction with the East Dal at Djursås (Djura) on one side of the river and Gagnef on the other; they followed back the

East Dal to Ahl, or Ål, on the Innsjö, and there took the high road to Falun. The last date in the journal is August 17. The mountains of Dalecarlia had now been twice explored fully on both sides. They and their leader had triumphed over many secrets 'wrung from nature's close reserve.' 'At the end of life *La Gloria* is sung,' quoth Näsman.

Linnæus gives a list of the parish priests who entertained them. This in itself incidentally shed a valuable light over the province; the priests could not fail of being illumined by his conversation. Linnæus always held that priests should study botany, for use in medicine in their out-of-the-way parishes, and science to enliven their own solitude; for use, besides, in their glebe, said Näsman, for ' good husbandry is good divinity.'

'Good courage breaks ill-luck to pieces,' was his final proverb. The whole party arrived safe and well at Falun after a journey of over seventy-nine Swedish (or 553 English) miles by the direct road.[1] To this

[1] *Distances in Swedish ¼ miles.*

Fahlun	Biursås	7
	Rättvik	12
	Oret	10
	Orsa	13
	Mora	5½
	Elfdahl	14
Särna	Serna	30
	Idre	16
Fielten	Fiellen	6
Norwegian frontier	Norvige	19
	Rörås	20
	Grüfum	4

we must probably add one-third for circuits and explorations.

Linnæus put into the governor's hands the journal he had directed and superintended of all the observations made on the journey;[1] at the same time he delivered to the Baroness Reuterholm's care her two sons in blooming health and high spirits.

Linnæus remained on his return from this journey at Falun, where he established a little college under the auspices of Baron Reuterholm, giving lectures on assaying and mineralogy. In a remote town like Falun the novelty of these instructions excited interest. He became rich, so to speak, in friends and money.

'Returned from my journey,' says Linnæus later in a letter to Haller,[2] 'I took up my residence at Falun, the capital of Dalecarlia, began to give lectures on mine-

	Rörås	4
Swedish frontier	Sverige	19
	Tokogne	12
Transstrand	Serna	16
	Lima	32
	Malung	12
Appelbo	Nås	24
Terna	Floda	8
Gagnef	Gagne	8
Gagneahl	Ahl	4
	Fahlun	12

Linnæus wrongly reckons this total as 313
It is really 307½.

[1] The *Flora Dalecarlia* has been lately edited by Dr. Ewald Ährling of Arboga.—*Encycl. Brit.*, ninth edition.

[2] Dated Stockholm, Sept. 12, 1739.

ralogy and was universally beloved.' In his diary he says, 'Linnæus here at Falun found himself in quite a new world, where everybody loved and assisted him, and he acquired considerable medical practice.'

He had now completed his mineralogical system, and read it, greatly to the satisfaction of the miners. Brouwallius, at that time chaplain to the Governor Reuterholm, and tutor to his children,[1] conceived a particular regard for Linnæus, and wished to be taught by him the art of assaying, mineralogy, botany, &c. Linnæus therefore began a course of lectures on assaying at Falun, and for this purpose obtained permission to make use of the laboratory belonging to the mine district. A considerable audience attended him.

One thing marred his good fortune. On his return from Norway he found the sad news awaiting him of his mother's death, at the early age of forty-four. This was a heavy stroke to him, and all the more so that he had not been a comfort to her—that his career had hitherto been a source of continued disappointment to her, or at best of hope deferred.

Now that he was beginning to flourish she was dead and could not enjoy the knowledge of his rising fame. One can only have one mother. He wore deeper mourning for her in his heart than on his person—wore it with a repentant, prayerful feeling like Johnson's 'Forgive me whatever I have done unkindly to my mother, and whatever I have omitted to do kindly.'

[1] Diary.

The memory of all her unrepaid kindness to him was a sad yet loved relic. Let him now pay her the honour she would most care for—in making her the mother of a distinguished son.

The softened frame of mind into which the tender memory of his loss threw him made him the more susceptible to impressions of feminine grace and beauty, and gave him a longing for woman's more intimately sympathetic companionship. 'His friend Brouwallius, afterwards professor and Bishop of Åbo,[1] saw no means of his getting forward in the world without going abroad and taking a doctor's degree, in which case he could, on his return, settle where he chose with advantage; and as money was necessary for all this, there seemed to his friend to be no alternative but for Linnæus to pay his addresses to some young lady of fortune, whom he might render as happy as she might render him. Linnæus approved theoretically of this advice; but notwithstanding several plans were proposed, no one was just then adopted.' Linnæus, in the previously quoted letter to Haller, speaks of his first and only serious love affair. 'The physician of that district' [Falun] 'passed for a rich man. Considering the poverty of the province, he could justly be deemed opulent. His name was Moræus, eminent for his learning and skill among the Swedish physicians. Physic, especially practical medicine, was the science which he esteemed and preferred above all others. He grew fond of me. I visited

[1] Diary.

him frequently, and always met with an amicable reception. He had two daughters. Sara Elizabeth, the elder one, was a beautiful girl. A certain baron had paid his addresses to her, though without success. I saw her, was amazed, smitten, and fell in love. My caresses and representations won her heart. She promised her consent and vowed to be mine. But, as a poor man, I was much perplexed to ask her of her father. At last I ventured. Moræus consented—and refused. He loved me, but not my uncertain and adverse fate. He finally declared that his daughter should remain unmarried three years longer, and at the expiration of that time he would give his ultimate decision.'

The diary gives the account of this affair with a slight variation. For both we have Linnæus's own authority. In one place he is speaking to an acquaintance, in the other to himself, or to posterity.

'Dr. John Moræus, physician of the town, who was looked up to as a man of considerable fortune (for his situation in life), and who saw the progress of Linnæus both with astonishment and jealousy, had determined never to bring up any one of his children to the practice of medicine. Linnæus, however, in spite of all this, and though a mere student, after having spoken to the eldest daughter, presented himself to her father, and asked his consent to marry her, which Moræus, to the great surprise not only of Linnæus but of others, agreed to; however, he could not obtain the consent of her mother.'

'*Voluit et noluit*,' he would, and yet he would not, writes the impatient lover to a friend of Dr. Moræus. Himself a successful doctor, Moræus could not bear the notion of marrying his daughter to a man of science, in its broad sense, without any fixed and definite line of practice. He counselled Linnæus to take the degree of doctor of medicine, which necessarily involved his going abroad, as at that epoch the university of Upsala granted no degrees to her own students.[1] Swedish students at that time used to graduate in some foreign university, and, like most of his countrymen, Linnæus fixed upon Harderwyk in Holland, as the cheapest place. But cheap as it was, he had not the capital to sink in the preliminary expenses, including the journey thither. He was in possession of thirty-six golden ducats,[2] earned and saved. His Wrede pension amounted to sixty dollars,[3] reckoned at 5*l*., per

[1] The university of Upsala does not (now) confer degrees unless the recipient of the honour has proved his capacity by passing a searching examination, no exception being made in the stringent enforcement of this wise regulation.—DU CHAILLU.

[2] Linnæus speaks of having 36 nummi aurei—meaning ducats, the usual gold currency of Sweden. Nummus aureus, the single ducat, nearly as large as a sovereign, but thinner and lighter than our 10*s*. piece; value 9*s*. of our money. It weighs 54 grs. troy; our half-sovereign weighs 60 grs. The Swedish gold coins were double, single, and half-ducats. 'Exivi patriâ triginta sex nummis aureis dives' are Linnæus's own words.

[3] Charles XII. struck large pieces of copper and called them copper dollars. They went down immediately in value. This causes great difficulty in reckoning the money values of that century. The large copper pieces in general, but more particularly the two-dahler pieces, are called plates (plåtar).

annum. He strained every nerve to obtain a continuance of this, but failed. 'He had enjoyed it one year, but after that nothing, and as soon as he went abroad he lost this exhibition through his enemies.'[1]

Falun had materially altered in its aspect for Linnæus since he had been absent. Sara Elizabeth, the elder of the two handsome daughters of Dr. Moræus, had come from Sveden, her father's country seat at some distance from Falun, and she, like the rest of the world in the Dalecarlian capital, was curious to see the interesting traveller who had recently returned successful at the head of an adventurous band of explorers. In fancy I can see their introduction to each other; they first shook hands, then she bobbed a curtsey, and he lifted off his hat. This is the order of the usual salutation in Sweden. The little girls and young women always dip a curtsey to everyone in the company; even the youngest boys never omit to take off their hats separately to each person. 'I was struck when I first saw her,' writes Linnæus to his friend Haller, 'and felt my heart assailed by new sensations and anxieties. Nature is nature wherever you find it,' whether in the land of Romeo or of Linnæus. Elizabeth, too, seems at once to have felt the strange power of eyes made to discover truth; and here was a truth entirely new to him—that the charm of a beautiful maiden is the most exquisite thing in the world. He who had counselled the young men, his companions, to keep their

[1] Diary.

heads free of love—was science all-sufficient for him now?

In the fact of Elizabeth Moræa's habitual residence at her father's country house is the simple solution of what has been a difficulty to Linnæus's biographers. In a small place like Falun—a town not then a hundred years old,[1] and not then mustering half its present number of 7,000 inhabitants—it must otherwise have been difficult for the Beauty of the place, called the 'Fair Flower of Falun,' to remain unknown to a handsome and accomplished young man whose society was sought by all the leading men, her father included, and who went about investigating, studying, and making himself known in all the separate villages which then composed the town of Falun.

Sara Elizabeth had never seen anyone like Linnæus before; he ousted the 'certain baron' completely from his place in her fancy. His ideas were an outlook from the sordid, mercenary views of the mere speculators in copper with whom she was acquainted, her father at the head of them for shrewdness and tightness in grasping. She could not help admiring this bright being who despised the gods of her family—cash and comfort—who showed her fresh forms of wealth surrounding her, and beauty even here in barren, blighted Falun; and he loved her partly for her beauty, but chiefly for her sympathy. He loved her because, as Mahomet said of Fatima, 'She believed in me when

[1] It was founded by Queen Kristina in 1645.

none else would.' Her father could not be expected to do the same; much as he liked Linnæus personally, it was difficult to win him over to accept the student as a suitor for his daughter. But finally, after infinite persuasion, it was arranged that if Linnæus would take a doctor's degree, and if he could succeed in making a sufficient fortune during a three years' probation in absence abroad, at the end of that time he should be accepted as a son-in-law, and settle down in Falun as a practical physician under Dr. Moræus.

So closely was Carl wrapped in the romance of love's young dream that he did not feel this to be a clipping of his wings, nor suspect that in this way the parents might calculate on being fairly rid of him, and the baron might come on again undisturbed. Hard as the conditions were, with his usual impetuosity Carl accepted them at once, and, to the astonishment of Dr. Moræus and his wife, he undertook to set out at once for Harderwyk to fulfil the first article—the obtaining of the doctor's degree. They would have been still more surprised had they known that their daughter, their prudent Elizabeth, brought him the savings of her pocket-money, one hundred dollars,[1] and gave them to him. How exquisite this Elizabeth was, thus to further his views, and make more precious the sacrifice she was ready to make for him, inspiring him with emotions of gladness and gratitude too deep for any words! Elizabeth,

---

[1] Stoever considers this sum, with his own thirty-six ducats, made about six hundred copper dollars.

as she handed him the purse (that bore Linnæus's fortunes), pleasantly cautioned him to be careful of his money. This care for him sounded divine: such playful lectures are sweet from the lips of a lovely girl who is beloved.

In early life Linnæus acquired habits of very strict economy and frugality—the habits only (in which poor people often seem to overvalue money); but the love of riches was not a passion with him, as has been untruly said. Indeed he was rather thoughtless in spending, and needed someone to manage his purse for him. Linnæus's one great passion hitherto had been for truth, for which in all respects he ever showed the most sacred regard. Elizabeth had been brought up to value money for its own sake.

She now equipped her lover in the spirit in which ladies sent forth the warrior knights of old. They kissed and parted.

> My own affections, laid to rest awhile,
> Will waken purified, subdued alone
> By all I have achieved. Till then—till then . . .
> *Paracelsus.*

Our philosopher went forth trustful, cheered, and stimulated, love-healthy, not love-sick. The words holy and healthy are both derived from the same old German word *heilig*. 'That old etymology, what a lesson it is against certain gloomy, austere, ascetic people!'[1]

Having spent the winter months in visiting his

---

[1] Carlyle.

friends and relations,[1] in preparing his academical dissertations, and arranging the collections of his materials of reform, which he considered his most valuable treasures, Linnæus at the age of twenty-eight **began, in April 1735, his** travels in foreign **countries**—the *Wanderjahr* which ' ancient custom rendered **necessary,** and which became pleasant by the happy prospects of his further improvement and the enterprises he had planned.' He was not one of those who wait to do great things till they are rich and have time.

It seems **from** what Stoever says, that Linnæus could not resist making a hurried farewell visit to Falun. I **fancy Elizabeth** forgave him this pardonable misuse of **some of her copper dollars.**

[1] So says Pulteney, but as **Stoever** asserts he left Falun in April, and as we know he visited Stenbrohult in his southward journey, it is most probable he spent the winter in Upsala, as he went there to look after his Wrede pension, which, he says, his enemies caused **him** to lose.

CHAPTER X.

TAKES HIS DOCTOR'S DEGREE IN HOLLAND.

> *Enter* ALMA MATER *as a Hag.*
>
> 'Powers!' she cried, with hoarse devotion,
> 'Give my son the clearest notion
> How to compass sure promotion
>   And take care of Number One.
>
> 'Let his college course be pleasant,
> Let him ever, as at present,
> Seem to have read what he hasn't,
>   And to do what can't be done.
>
> 'Of the philosophic spirit
> Richly may my son inherit;
> As for Poetry, inter it
>   With the myths of other days.
>
> 'Cut the thing entirely, lest yon
> College Don should put the question,
> Why not stick to what you're best on?
>   Mathematics always pays.'
>
>   JAMES CLERK MAXWELL.

THE tree of all others that our botanist had now to study was the money-tree—how to make that bear, or rather how to sow his 600 copper dollars (about 15*l*. English, according to Smith) so as to yield the best return.

Linnæus left Falun in April 1735, some say in company with Claudius, or Claes, Sohlberg, one of his earliest Upsala pupils, the botanist of the Dalecarlian journey, a medical student also proceeding to Harderwyk to take his degree; but as Claes Sohlberg was now at Lund, Linnæus would more probably have met him there or on the coast. We do not hear if he had as many or more copper dollars than Linnæus, but he does not seem to have been a vastly interesting companion; at any rate no kindred soul like Artedi. But Artedi was in England trying to transmute his learning into gold. Perhaps he and Carl hoped to meet somewhere abroad.

Carl sailed down the rough blue Lake Vetter to Jönköping. He did not linger to enjoy the pleasant promenades of Jönköping by the lake, nor to ascend Dunkellar Hill with its beautiful views, where now are numerous villas with well-planted gardens, testifying to the profits of the roofing-paper and match-making trades; nor was he tempted by the merely picturesque charms and waterfalls of Husquarna. But the famous iron-mountain of Taberg did not lie much out of his way—at least not in his mode of travelling, which was very frequently on foot. So leaving the high-road, which even at that date we may assume to have been a bridle-path, he bore away westward, following the course of the stream flowing from Taberg into the Vetter, and ascended Taberg at about eight English miles south of Jönköping, from which height, 1,096 feet,[1] he gained a grand survey of the forests of Småland,

[1] It is 1,129 feet above the sea.

and investigated the iron-mines, which had a special interest for him after his mining studies in Dalecarlia. This celebrated iron-mountain, with a few others found in Lapland, are the only ones in Europe where the ore is broken or blasted above ground. Taberg was doubtless the attraction that determined his route to Holland this way; and also the wish to revisit his home, to which his memory always affectionately clung.

He travelled on southward by way of Eckersholm and Lindefors, following down the Laga stream to Wernano, where it expands itself into the fiord-like lake of Widastern. Wernano, on account of its well-attended annual fair, has good roads leading to it from all four quarters; but to ease his walking Linnæus took a boat down the Widastern beyond Berga, where the lake contracts as far as Ljungby. Here a walk of ten English miles through a familiar country brought him to Lake Möckeln, where he was at home, and where any fisherboy would gladly give him a lift to Stenbrohult.

His father was a lonely man now, being a widower; his two elder sisters were married, and his brother Samuel was away studying for the ministry. For all his mother's careful wish that he should not enter the garden, husbandry and natural science remained ever Samuel's favourite lore. He shone later as an author and one learned in entomology, which, he flattered his conscience, was a branch of farming, in his heart classifying farming as a branch of entomology. Carl felt a renewal of his sadness at not having been able to see his mother before her death, but his letters had arrived so long after the event. Distances, too, are so huge in Sweden, and poverty forbids the indulgence of grief, a luxury permitted only to the rich; but revisiting the scenes of his childhood brought his mother doubly near to him.

'His mother was not permitted to see the successes and honours which her eldest-born was destined to achieve. Poor mother! Her sun had gone down when it was yet midday; she had borne the burden and heat of the noon, but the season of rest, of in-gathering and rejoicing she tasted not in this life, for she is laid in her lowly grave in the shadow of the church of Stenbrohult, and thither her son repairs to shed in secret the tear of filial love and regret. Perhaps he has never more longed for the sympathy of a mother's heart than now, when he feels the anxieties and fears of hope deferred—and to whom could he have so unreservedly communicated the thousand hopes, joys, fancies, and desires that throng around his heart as to her who lies there? Ah! in vain he sighs and longs for some response; there is no sound save that of the murmuring breeze that waves the harebells which cluster over the green sod beneath which she lies. "Alas, my mother!" and again, "Alas, alas, my mother!" he cries, and bitter tears fall fast. But soon he has dried them; he may not yield longer to grief; the day of life is yet before him, and he must gird himself and go on his way, and do his work ere it be night, and he too shall lie down and sleep.'[1]

He walked over to Wirestad, where his sister, the lively young Anna Maria, was petted and beschoolmastered by her Gabriel in a tender and clerical way. From Stenbrohult the forest road to Traheryd[2] again follows the course of the Laga to Markaryd, where Carl had cousins. Here the river flows off westward to the Kattegat and the bridle-path works southward, only turning off westward at an elbow when within a few miles of Qvidinge,[3] where there are recently discovered and extensive coal-fields. Had Linnæus any suspicion of these valuable stores lying beneath his feet? Fame and fortune were both lying unrecognised beneath his tread.

Modern power, stored in the coal, and mediæval power as represented by the ancient picturesque Herrebadskloster, meet on this spot, and several railway-lines now open up this district to common knowledge. Coal-mines and a popular bathing-place bring money into this corner of Sweden. Of Helsingborg Castle—grey ancestor of the town—on the hill, only a picturesque ruined watch-tower now remains. The tower of the old church is even more delightful, its scaled gable rising with lanceolate openings at each step. Here it is most likely Carl met his travelling companion, as Helsingborg

---

[1] Brightwell.

[2] Which must not be confounded with Traveryd near Lake Bålången, where the floating islands are.

[3] Where the Crown Prince Carl August died suddenly in 1810 and changed the dynasty.

is at no great distance from Lund; and in less than half an hour the two students are in another country.

Elsinore had no especial attraction for Linnæus; he cared nought for the traditions of Holgar the Dane, and he had not read Shakespere. It is doubtful if 'Hamlet' had then been translated into Swedish. It is translated into Danish now—I am told by Norwegians, who think it clever (!), that many readers hold it even second to Ibsen. Hamlet's and Ophelia's graves had not then been invented.

These tombs were unknown to Hans Andersen, who lived at Elsinore—not in his childhood, I think, though Hare says so, but when he was nineteen and studying with the rector of Marienlyst. The Kronberg has really little to do with any poetical prince of Denmark, for 'Hamlet had in fact no especial connection with Elsinore: he was the son of a Jutland pirate in the insignificant island of Mors.'[1] Still the platform of the Kronberg will be famous as long as Danish history lasts; and no Danish history will be accepted as complete without Shakespere's 'Hamlet' and his woful tale.

Carl's journey hence, probably the easiest he ever took, is the most difficult of any to his biographer. Turton says he travelled across Denmark to Hamburg. Stoever of Altona, his personal friend in later life, who had abundant means of knowing all about his hero's life in these parts, says he went through Jutland, Schleswig, and Holstein; most writers say he visited Copenhagen;

[1] Hare.

and as Linnæus says he did so at one time of his life, there seems but one other time when he had the opportunity, which was when he travelled in Skåne, the sixth and last of his tours. Linnæus's own diary says 'he continued his journey through Helsingborg to Elsinore, thence by sea to Travemunde and Lübeck, thence to Hamburg,' which seems clear and positive; but Pulteney, who edits the diary, coolly sets it aside, saying he passed through part of Denmark. One can hardly call touching at Elsinore travelling in Denmark. Indeed Linnæus himself mentions elsewhere (and previously to the Skåne journey) that he travelled in Denmark. There is something rotten in the state of biography when discrepancies like these exist. We often find Linnæus in far-off places. Where he dropped from we cannot always ascertain. We make light of this in cursory reading, but in actual travel the distances are enormous—much too long to hop or skip. It is, as Fuller remarks, 'easie for a writer with one word of his pen to send an apostle many miles by land and leagues by sea, into a country wherein otherwise he never set his footing.'[1]

Either view of the biographers gives Linnæus a strangely roundabout and expensive journey to Hamburg, where they all meet on neutral ground. For my part, I believe Linnæus's statement in all its simplicity, considering the scarcity of copper dollars; but as written history is so positive the other way, I have constructed a neat little hypothesis to meet their views, which has

[1] The *Church Historie of Britain*.

the merit of reconciling all their statements. It is too plausible to be true, and I do not think he took this line, unless, indeed, he went to look for some little weed in its native habitat; but as George Eliot says, 'I am beginning to lose respect for the petty acumen that sees difficulties.' He says he went by sea from Elsinore; well, he took ship from Elsinore to Aarhus, the capital of Jutland, where Pulteney finds him; thence he went by land to Neustadt, where Stoever picks him up, and by boat to Travemunde and Lübeck, where truth, history, and all the biographers make sure of him. He made the journey through the north of Jutland on his return, as he went by sea from Rouen to the Cattegat, where the wind went contrary and his farther voyage that way was stopped; therefore I urge that he travelled from Fredrikshaven by land to Aarhus, whence he returned, as we know he did, to Elsinore. And this is the route I took in order to follow him according to my hypothecated plan of his travels. I speak as *Chorus*, and shall describe the country as I saw it.

Aarhus is a large town with fine streets—one of them raised on a viaduct after the manner of our Holborn—several handsome churches with *carillons*, and an old romanesque brick cathedral with a remarkable tower. This church is frescoed and painted elegantly inside. Here the two medical travellers landed, we must suppose, and proceeded southward with all speed. The evening mists hung heavy on the fields, and the climate felt milder and moister than in Sweden. More barley than rye is grown here, and at Aarhus one meets with vegetables cheap and in plenty. I have bought for next to nothing the juiciest radishes I ever ate; when at Christiania, they were tough as fir-cones, and at famine price. It feels warm, but there are snow-scoopers to all the trains; it looks peaceful, but the Danish camp is arrayed with banners before a large square castle, built in the aggravated-barn style. The black and white cross-barred thatched houses of the peasantry look more comfortable, and group prettily into villages, among which the storks move about with stately familiarity.

How soon one sees proofs of the greater prosperity, in actual coin, of the people here over the Northern Scandinavians! They wear much more jewellery—if the Danish filigree can be called jewellery—than the further Hyperboreans wear: many more rings and bracelets—trash, if you will, and these things look like it—*messing* is their name for it; but even trash shows money to spare, and all appearance breathes of Jutland's monotonous fertility and middle-class ease; not amounting to wealth, with its splendour, its picturesqueness and wickedness. This is such a virtuous-looking land, dull with the decorous domestic virtues, upon which, however, a fine building, or a fine character, shows to the greatest advantage. The people are withal as fearful as the Germans of fresh air, even now in June. The lake scenery of northern Scandinavia is continued through Denmark, though the banks, set in sunny colza meadows, are softer and more sleepy. But here the fir trees are

Christmas trees, in Norway and Sweden they are pines. We often read of richly timbered Denmark, and here are some wind-tossed oaks of small size; but in all this region I have not seen a beech tree of respectable girth. Jähns, speaking of the pre-historic ages, says, 'In Denmark their division is marked even by the vegetation. The stone age lies buried under the fir trees, the oak stratum conceals the bronzes, and the iron age is covered by the birch and elders'—a captivating idea, though I don't quite see the drift of it, or I would begin to dig at once. The four sorts of trees are all here closely side by side. It is one of those deep German ideas that mean anything you please.

Here is a pretty fiord near Veile, with bulrushes in the foreground, set in undulating country covered with golden broom, backed by beech-clothed hills; but these are woods, not forests. When people talk admiringly of the beech forests of Denmark, it is more often Holstein they are thinking of. Here is Fredericia on the belt of sea dividing Jutland from Funen. The strait (a silver streak) and a weed-fringed foreground stream winding among the black peat-pits, give light to a pretty Danish picture. The peat-pit is everybody's own private coal-cellar. Beyond another fiord, with a pretty hamlet on its border, one sees Kolding with its high-placed ruined castle, of which one battlemented tower remains solid, while all the rest is gutted—a mere shell. The canal now makes the timber-station of more importance than the proud castle, which stands all solitary but for the heavy heron's flapping flight. Here we cross the borders of what is now part of Germany. Slesvig was Danish in Linnæus's time. Here are women working in the fields—a sight seldom seen in Sweden. The yellowing barley and blue-green waving blades of wheat have superseded the browner blooming rye of the north. Here, too, are thistles—a rarity to Linnæus, there are so few thistles in his country. Here they fix the heads of the cows and sheep in a long heavy wooden framework to keep them from migrating over the border. It is a most uncomfortable form of collar.[1] There are goats (collarless) and various other animals about. The ground is level here and boggy; not so well farmed as Jutland, nor so well peopled; a dismal, swampy, flat, unprofitable country. I marvel at the Germans coveting it. They did not need it even to complete their ring fence. Perhaps if they had left it under the government that does not require so many soldiers—'Death's staunch purveyors'—it would have been better for the land. 'One sighs to think how these unproductive consumers of Wurst, with all their blue and scarlet broadcloth, are maintained out of the pockets of the community.[2]'

The country is all but emptied for military service; otherwise it might be good land if cultivated. We hear nothing of compulsory military service for the young men in Linnæus's time—in Scandinavia, that is. The land improves as we travel on, but it is still flat as

---

[1] What is their Society for the Prevention of Cruelty to Animals about? [2] G. Eliot.

a board, and with no woods. The towns crowd about the broken ground where it rises anywhere; one is so glad of a rising line as a relief to the eye. One is scarcely satisfied by the molehills nature here provides as a substitute for mountains. Man, feeling a need for more upright lines, built spiry steeples about here. We find this refreshment to the eyes in pretty Flensborg, with its spire set off by four small round spirelets clustered at the tapering point, its tiled gables and general diversity of level, its gardens and roods of yellow water-lilies, and its fiord, which is not visible from the railway. There is a small branch line into Flensborg. Friends gather in numbers on the platform to enact the moving scenes of real life and to wave the tear-bedewed (?) handkerchief at starting. It is quite a lounge for the town. Outside we find haymaking going on, and, for objects of interest, oats and white lambs, men in white ducks, broad-beans in flower, and pease-blossom, moth, and mustard-seed.

How summer crowds upon us in this journey down! the season changes so rapidly. Beyond a beech plantation is some rising ground, and here is a *railway cutting* at last! A great square barrack of a castle looms to the left, set in undulations and green hedges, and the mixed cultivation so refreshing after the monotonous vegetation of the north, and kitchen-gardening beautified with purple cabbages.

Schleswig on the Schlei—a scattered town with spired churches cresting banks of pink campion and genista in full blaze. I do not see why Linnæus should have been so astounded at the furze-gold on Putney Heath. This yellow broom is just as full of sunshine.

I am bringing you down express through the country, that I may efface myself the sooner: to be personally conducted is a bore.

Level again; the undulations sweep away to the westward. Many fine bay horses are grazing in this part of the country. Another river, or fiord rather, is overwaved by tall ash trees at Rendsborg, whose Dutch aspect shows how far southward we have moved; the canals reflecting its brick church with lance-like spire, and its Gothic and modern Gothic buildings; the level land melting off blue in the distance on one hand, and dimpling into dune-like hillocks on the other.

Hedges again, and wild roses, and girls binding

> June lilies into sheaves
> To deck the bridge-side chapel, dropping leaves
> Soiled by their own loose gold meal.—*Sordello*.

After all one misses a good deal by living in the North. One misses summer; it is all spring or autumn there. My closer calculation is that in Sweden June is spring, July summer, August autumn, and all the rest winter. One is glad to see rich and varied vegetation again, and roses twining over porches, and to smell the sweet syringa. The hay, the flowers, the breath of the whole land comes up in great wafts of fragrance—a scent of summer bloom, of clumps of pinks and picotees, and tall fiery tiger lilies, with cows licking each other all so friendly in the shade of great umbels of elder, such as must have made Linnæus think of his former comrade Artedi, now gone to England. This sudden drop into summer is very striking in coming from the North, and must have been especially so to the born Swede Linnæus. It is like the descent into Italy from the Alps, or into Damascus from Mount Hermon

whitened with eternal sleet,
While summer in a vale of flowers lies sleeping rosy at her feet.

There is less length of daylight to see it by, however: the sun sets three half-hours earlier here. One is startled to see the distance purpling under the golden light, with sunset reflected in the eastern banks of rose-tipped pearly clouds; and to feel the humid airs of grateful evening mild distilling from the greenery of fat and fertile Holstein—a land of plenty and prosperity, with hedges deep in barley and long grass to mow, and yonder towers and spires of Kiel all grey under the veiled misty golden moon.

The blossoming gardens here are turned to fairy-land, as it would have seemed to Linnæus; to us it is more like the opera stage, with groups of girls, with faces more of the pretty Danish type than like the plainer Germans, dressed in short full skirts of plum-coloured velveteen, or of dark stuff with a coloured velvet band round the hem, tight-fitting basque bodices of black velveteen with very short puffed sleeves, showing the arm bare high above the elbow, and white bib aprons—altogether absurdly like the theatre, and quite as pretty.

A tiny white net cap, the size and shape of a pork sausage, made of a quilling of net round five inches' length of lace, is stuck on at the back of the head. This costume is generally worn by maidens of the industrious orders in this part of the world, and very neat and maidenly it is.

Timber-wharves show the speciality of Kiel, besides the men-of-war laid up in the harbour, and the numerous officers and seamen, whose capping of each other and the military reminds an Englishman of Portsmouth, as he walks along Kiel Strand.

Morning sunshine shows the lake scenery of Holstein to advantage. It is very pretty country from Preetz to Greemsmühlen with their fine beech woods encircling their lakes. But Eutin seen in a heavy rainfall looks like a failure, which is disappointing, as Bædeker praises it so highly. The sea-gulls only can venture out; but the varied landscape of hill and dale, with streams and happy thatched homesteads, looks pleasing in the worst of weather, as one sits in a sheltered window eating white bread and the white butter given by the brown cows feeding in the black bog-beplastered fields. The people look rosier and prettier, and in fact more Swedish, about here than elsewhere in Germany; so Linnæus was the less pained by the contrast. Our hero went by sea from Neustadt to Travemunde, thence up the sleepy river Trave to Lübeck. Here we are certainly on his footsteps. Stoever, Turton, Pulteney, Fée, and the 'Hamburg Courant' all agree with Linnæus himself in this.

The duodecimo biographers who never think, but only copy, of course agree likewise. *Exit* Chorus.

A spiry place is peeping above yonder hill. It must be Lübeck. It is so. Now it looks like a dozen or so of spear-pointed steel pens with their points upwards as our two Swedes arrive at the bridge by the gate of this quaint and ancient city; an old, and what young persons who sketch call a most rapturous, gateway, dated 1477, bent with its age and weight, with a sheaf of spire-like lances rising behind it—the shield and spears of the city. Above the portal arch is the legend CONCORDIA DOMI FORIS PAX. This heavy pile is shaped into two round towers, cone-spired, with a burgomaster's house betwixt them, from under which the broad low portal has been hollowed. The gate is brick-built in stripes of black and red, the black bricks, being vitrified, look white and lustrous at a distance, giving altogether a peculiar effect as of a large mineralogical specimen. It is all sunken in the ground, which has been excavated to leave its roadway clear.

Friends were busy at the wharfs and coach-offices meeting friends; but no one knew Linnæus and his companion, bewildered with their first experience of encountering an entirely foreign language. The two Swedes carried their chattels up the hilly street, where now the tramcars climb; and, oh! they did indeed find views and architectural surprises. Linnæus was as yet a novice in architectural criticism, and I find no comments so far in his diaries. It was all what he might have expected in the South. But even more experienced travellers than he have never seen any town so odd, so quaint, so Gothic, so riotous in invention, so queerly coloured, and so charming. Red-brick buildings, shiny black-brick arcades, and Gothic frontages, all crying out 'No dinner for you to-night; so long as twilight lasts come out and look at me'; Renaissance and florid architecture of all styles retiring only to await their turn for attention. Other Gothic towns, even Bremen, subside into commonplace before the luxuriance of Lübeck. The shields on the buildings are gilt, coloured, and slung sideways—anything for extravagant effect. Here are spires of all sizes, and gable-forms from great A to little z, and as the city lies on a slope they can all be seen together, like the diagram at the South Kensington Museum of the typical buildings of the world and their relative heights, from Cheops' pyramid down to a drinking-fountain. The drapers in this part of Germany have a way of arranging the basement cellar window beneath the shop in the same manner as the shop above. In Denmark the cellar window is also dressed, but it usually belongs to a different establishment, and generally vends vegetables or beer.

The best way to see Lübeck is to do as Linnæus most probably did, as he was a great one for hill-climbing—that is, to take an evening walk up a forest-covered hill called Chimborasso, just outside the town, and so gain a concrete idea of the place, that one may the better relish afterwards the full and bewildering effect of all its oddity and **bustle, and** realise the old-world scene **in** full perfection in the morning when the marketplaces are overflowing and streaming with colour and picturesque confusion. On the hill-top is erected a sort of scaffold staircase with benches, whence one can enjoy the view. The prospect from Chimborasso is of a mass of bright red roofs on the opposite hill, individualised by the white-gabled window-dotted fronts of houses, set in a forest of leafage melting into **blue** distant meadows, where two rivers—the Trave and its tributary —meet as they mingle with the verdure. The spires and pinnacles rise out from among the foliage and the town-roofs to the sky. Plenty of life and **activity** is going on below, its buzz mingled with murmur of windmills and sounds of trains and carts, of birds, and bells, and boats.

Now we may recross the river to the town, ascending the hill, first on the side of the well-nigh deserted cathedral, which is always under repair, and always needs it. Its tall thin spires are so much out of the perpendicular that they would look atrocious in a drawing; for one's reputation's sake one must straighten them and make them more untruly true. To stand under these spires **is one** of the risks of travel. It feels less dangerous inside, where one does not see them. There are some remains of early coloured glass by the painter of the cathedral windows of Florence—'that most singular master who, in this art, was known in the world,' who was brought for this purpose 'from distant Lübeck on the Northern seas.'

Most of the monuments take the form of portraits, chiefly with ruffs round the neck, and many of them well painted. There are some pictures of the old German school also, and an interesting medley of ornaments disposed about—a kind of ecclesiastical trinkets. There is a quaint clock set on an inner rood-screen of carved wood with coloured figures, of which those above the clock strike the hours. Death turns his hour-glass upside down, strikes on the bell the number of the hour, and shakes his head. Life, a figure in green, strikes the half-hours and the warning before the hours. The face in the centre rolls its eyes at every tick. These old mechanical clocks are among the few things belonging to the towns that Linnæus remarks in his diary. The clock in St. Mary's sumptuous church is still more elaborate than this, and the market-people crowd in to see the procession of figures as the clock strikes twelve at noon. The wealthy burghers of Lübeck meant to outdo the cathedral by building a finer church, with finer monuments of themselves. St. Mary's Church is bewilderingly rich in monuments in gold and colour, and Renaissance ornaments. The whole church seems a co-operative monument of family pride.

The market-place on market morning ranks high among the sights of Lübeck. The square and the surrounding colonnades are all crowded. The market-women wear straw hats tilted like bonnets, set on hind part before, with a broad ribbon of bright colour hanging behind in a loop. The hat fits over the little sausage-shaped cap, which sometimes has two narrow bands across it. The fishwives have blue ribbons on their hats, the fruit and vegetable sellers wear green, or both occasionally wear purple. They, and their customers likewise, carry pairs of baskets of vegetables, each a yard long, hung horizontally from a yoke. The butchers' market is held under the longest of the pointed-arched colonnades that divide the market-place from the busy interesting High Street and from St. Mary's Place.

From Lübeck Linnæus went to Hamburg, a city at least as ancient as Lübeck: it is said to have been founded by Charlemagne, but, oh, how different from Lübeck! a splendid great modern republic of a town, interesting to the student of social economy, made by commerce for commerce only, with its two different quarters inhabited by the masters and servants of commerce. The landscape leading to it from Lübeck is more varied in type than the rest of the vast monotonous plain of Northern Europe, and more happy-looking, having here and there a turfy slope covered with beech trees.

In Hamburg the market-women wear very short ungraceful petticoats, and their black hats have stiff black buckram bows at the back beneath the brim. Strawberries come in early here, and roses. This is a land of summer. Sweden is the land of spring, and Norway the land of winter. The overhanging houses in the maritime and commercial quarters of Hamburg are gabled and quaint, tile-roofed and cross-patterned with wood, looking old and very German. The tall church towers, bulbous or spiry, most of them copper-sheathed, and green with oxide, stand boldly up above the great city. The rich residential part of Hamburg by the Alster Bassin is handsome, new, and Parisian. Stoever would be the safest guide to Linnæus's history here, as besides being an Altona man, he was acquainted with Gieseke and the newspaper-writers, and all the persons interested in Linnæus's stay in Hamburg, where he created a sensation of rather a Mr. Verdant Green sort. But Carl's own diary is quite full enough here, and more moderate in tone than one would expect of a greenhorn.

Professor Kohl and Dr. Jänisch, and the licentiate-in-law Sprekelsen, who held correspondence with the best naturalists and botanists of the age, and who had a beautiful garden, all showed him great civilities.[1] Here Linnæus employed his whole time inspecting the fine gardens, museums, and everything else worthy of attention. 'It was like coming suddenly into a large inheritance of unknown treasures': he had to settle down to enjoy his new property. In a private library at Hamburg he found the botanical work of Ray, which he had long wished to see.[2] Among other things he was shown the museum of the burgomaster Andersson, and a monster which Hamburg gloried in possessing—the famous *Siren lacertina*, or stuffed Hydra with seven heads, belonging to Andersson's brother. This rare masterpiece of nature had formerly been exhibited on an altar

---

[1] Diary. [2] Brightwell.

in a church in Prague.[1] 'Linnæus was the first person who discovered that this wonder was not a work of nature but of art.'[2] Of course he proclaimed it. The price of the monster at once fell from 750*l.* to less than 5*l.* Carl examined the heads and found in them the jawbones of seven weasels. I dare say he crowed and was cock-a-hoop and overbearing like many young scientific men, of whom one prophesies that they will be charming at forty. The youngster never minds where he places his uncomfortable truths, nor whose idol he shatters. The owner of the Hydra, in true mercantile style, had fixed an enormous hypothetical value on the manufactured article. It was pledged to Andersson for the sum of ten thousand marks. The Anderssons were furious, an outcry was raised, and it was insisted on that the calumniator should prove in academical form, in public dispute, that the serpent was not a phenomenon, or that he should own his error.[3] Dr. Jänisch gave Linnæus the friendly advice to quit Hamburg with all possible speed to avoid litigation. He did so and avoided the discussion, leaving the Hydra master of the field. He often said afterwards, 'I had only one friend at Hamburg; this was Dr. Jänisch: he was a true friend to me.' Carl had made the place too hot to hold him during the month he stayed at Hamburg. The poor student could not compete with rich proprietors. The earthen pot would have been smashed to pieces by these heavy brazen pots. His vanity had made him

---

[1] Brightwell. [2] Diary. [3] Stoever.

over-display his learning in a place where his countrymen and their books had already the reputation of being too pedantic. He showed off, and they took him down a peg.

The object of his journey was as yet unachieved, and Carl's copper dollars were melting away: he had not been sufficiently careful of his money. We do not hear what Claes Sohlberg said to all this—whether he had gone on good-humouredly playing second fiddle to Linnæus, or if he had already proceeded to Harderwyk; but Linnæus took boat at Altona for Amsterdam. Shore and harbour were full of masts and spires, bristling with them. Linnæus had nowhere seen such wealth as here; it was marvellous to him when he first saw it. He had thought Upsala and Aarhus grand. What were these to Hamburg? Surely this must be the finest place in the world. No, he had read of Paris and of London. Should he ever see them? He had no previous notion of the greatness of the carrying trade. Here men and cattle and all sorts of goods embark and disembark, coming in from Harburg and all quarters to be shipped off to Holland and to England. Here is a fleet of fishing-smacks drying their nets, yonder are luggermen wetting their tan sails by sprinkling them with their oars. Linnæus takes the cheapest passage he can, and bargains hard for that—with difficulty, as he can speak neither German nor Dutch. The captain is his own pilot, and cautiously threads his way among the multifarious craft, and the ice-breaking vessels, with strong rams, laid up out of use beneath the Seamen's Home, close by where are now the fine hotels built on green pedestals of hillocks, where the sheaf of light-green bronze spires tower above the tall and narrow warehouses at Altona. The ship sails down the broad Elbe, past the fine suburb of Blankenese, where the merchant princes dwell, and out to sea by Heligoland. During this voyage Carl says 'he was exposed to great peril'; in other words, he had a rough passage. Stress of weather caused them to run in to Emden to avoid being driven on the shoals that are so numerous between the outer islands and the coast.

Emden is a pretty, picturesque town, and the 'Weissen Haus Inn' gives a good view of the town-hall, the river and bridges, and the prettily grouped and coloured houses. Emden is a thoroughly German place, where they know nothing whatever of English people. The beds are a droll experience. The top sheet, buttoned on the quilt, is only meant to reach halfway down; one sheet cut in two makes a pair. The rest of the bed is ultra-German in its manifold discomforts; but it is all so soft and downy that it feels like going to bed in a batter pudding.

At Amsterdam Linnæus stayed eight days, and saw all the splendour and expense bestowed on that city.[1] He then went by sea to Harderwyk, one of the dead cities of the Zuyder Zee, where, having undergone the requisite preliminary examinations and defended his inaugural thesis on the 'Causes of the Cold Intermittent Fever,'

[1] Diary.

he was admitted to the doctor's degree on the $\frac{13}{24}$th of June. I dare say the aspect of the ground in Holland and on the journey down added many ideas to his thesis, already prepared in Sweden. Linnæus had a keen eye for country and climate from the hygienic point of view. In the dedication of the 'Hypothesis Nova de Febrium Intermittentium Causâ,' among his Mæcenates and Patrones we find the names of Rudbeck, Rothman, Stobæus, Moræus, &c., and *Rosen*. The hypothesis here advanced, most correctly so denominated, is truly Boerhaavian.[1]

Here Linnæus won his doctor's hat, which is still preserved in his house at Hammarby, near Upsala. It is of greenish drab felt that once was green, turned up on three sides, with a pinkish bow that once was a red cockade. The university of Harderwyk, now long since swept away, was founded in 1372. In 1441 it contained more than three hundred foreign students. Harderwyk University gained a justly merited celebrity before it disappeared; Boerhaave and Linnæus, who have adorned the whole human race, graduated here.[2] The marble bust of Linnæus stands in a niche in a red-brick octagonal tower[3] standing by the site of the former botanic garden (not that of an ancient cloister, as Havard asserts), beneath a grove of tall trees which play Æolian harmonies at eventide as the inland sea breeze wakes the memories of Harderwyk's palmy days.

Harderwyk, a town of 5,000 inhabitants, now the

[1] Smith.   [2] Havard.   [3] Called the Linnæus Tower.

depôt and training-school for recruits in the Dutch East Indian army, styled an unruly and violent class of youths—they seemed quiet enough at the time I was there—was formerly 'the shepherds' refuge,' whence its name. When the Zuyder Zee extended itself beyond its actual limits, the wide meadows on its borders were sometimes suddenly flooded, and the shepherds with their flocks had to seek in higher ground a shelter against the encroaching waters. They built several huts, and were soon joined here by the fishermen. In 1229 the shepherds' refuge, become by the grace of Count Otho the town of Harderwyk, held rivalry with Hamburg.[1]

It has now chiefly a seafaring population. As one wanders by the shore of what here looks boundless as an ocean, the seamen of the one or two vessels which at most enter the little port together, larking among themselves, alone ruffle the tranquillity of the scene, a peaceful Dutch landscape of a low coastline, a few trees, roofs, and a little jetty set on the verge of an expanse of lustrous silver sea, flushed with the pink after-glow of day. Only an occasional fanfaron from the East Indian military depôt wakens faint melancholy echoes round the quiet shore. The turf near the sea is rosy-lilac with the thrift, as if reflecting the tender pink of the sky.

The academic quiet was as deep when Linnæus paced up and down here in thought. He had cause for thought if not for anxiety. He tells us himself:[2]

[1] Havard. [2] Diary.

'Now all the money he had carried with him from Sweden was expended, and being unwilling to trouble his father-in-law (that was to be), whose disposition he well knew on this score, he accompanied Claes Sohlberg from Harderwyk to Amsterdam.'

Thoughts of his doubtful future would obtrude themselves even during his eager study of the natural objects round him in a country where he found so much food for reflection. In Sweden he had seen how subordinate a part man plays in fashioning the appearance of the country, whereas at Amsterdam the mighty works of man's device—a miracle of human industry—had literally made the land. It was the reverse in this place. On looking at the wilds round the Zuyder Zee it was difficult to realise their possible transformation into a prosperous country. The magic wand of capital had never touched them. Yet there was hope even for these natural dykes and banks shielding the salt marshes and crossing the dull flats; for the ' good God was watching them as carefully as He did the pleasant hills inland; perhaps even more carefully, for the uplands He has completed and handed over to man that he may dress and keep them; but the tide flats below are still unfinished—dry land in process of creation, to which every tide is adding the elements of fertility.'[1]

And God would care for the student too, for he had it in him to be industrious and patient. Linnæus looked at the tender flowers; the thrift beneath his feet and

[1] Kingsley, *Glaucus*.

the orchis of the sands of Zeeland were consolation, ay, and wealth. Harderwyk calmed his mind—it was in harmony with his poverty; but he could not remain there making no money. His talent now confirmed by his doctor's hat, he must display it to the world and see what price the world would pay for the article it symbolised. It were easier to live upon the hat than upon talent such as he had displayed at Hamburg. After a breakfast of bread and raw eggs, as they love them in Holland, and coffee sipped from egg-shell china, such as the Dutch learnt early to encourage the making of, in imitation of the fine imported Oriental ware, the learned young Swedes set out on foot for Amsterdam.

Linnæus had already walked out Zwolle way, through a desolate country of white sand-hillocks with sparse fir trees, where he might well expect the vegetation to have a character of its own, as none but the fittest could survive. Here he admitted the truth of what a patriotic Dutchman told him, one who conceded that there are no mountains in Holland, but hills, he declared, there were in plenty. Here are, indeed, sand-hillocks where the dunes have spread in from the sea, extending for miles inland and along the coast of the Zuyder Zee. Inland the ground we travel over gradually rises, and is sprinkled with fir trees, brooms, and newly-set beeches. The land being in process of creation, the fir-crested dunes impinge upon the old chaos of black waste with dark tussocky grass, like evil heather; by degrees the sand will fill and dry the pits of bog and reclaim the gelatinous peat. The view is bounded by a range of dark purple hills—Dutch hills, that is—scantily clothed with firs and occasional beech and birch saplings, but no subsistence for man or beast. The land is not prepared to receive them yet. There is nobody moving about here, but further on we find a few old women wearing their silver heirlooms helmetwise upon their heads, and some 'mannikins' (small boys) watching that the birds do not make off with the occasional blades of barley. The land gradually gets less sterile; cattle, trees, and grass appear at Hatten, near which are flooded marshy meadows, and a bridge over a river, and Holland, as we know it best, appears again at Zwolle. But it is mostly a flat treeless country, with less capital laid out, and fewer inhabitants, than in West Holland, which is so much better situated for commerce. One looks out for the tumuli, or giants' graves, that one has heard of, but one only sees herons standing patient as monuments. The cottages are thatched, the few that exist, among the swamps and black peaty wastes, reminding an Englishman that the Frisians are his nearest relations. No wonder the Frisians and Saxons came to England; it is a vastly more tempting country to settle in. This is a desert of bog and sand, dotted with a few long-woolled sheep; the horizon is a dark indigo purple stripe, the middle distance a stripe of dead brown, the foreground a stripe of mottled drab; it is as dismal a country as one can see, with clouds lowering over it, few hands to labour, and no capital.

It shows what all Holland would be but for the capital supplied by commerce. At Krops Wolds the land improves into pasture; but these wolds and fens are no better farmed than ours were in the days of the Heptarchy. I do not suppose Linnæus wandered as far as Krops Wolds, perhaps not farther than Zwolle, of whose nine gates only the picturesque Sassenpoort remains.

All this was rambling for pleasure and research in the interval of waiting till he knew whether he were an M.D. or not. Now he had to travel in real earnest and for his life. Scarcely a copper dollar remained—he says not one; therefore we may conclude he had only some very small change. The Swedes packed their papers and trophies, slung their new showy green-and-pink hats, put on their old ones for use, and tramped to Amsterdam.

'Men think to mend their condition by a change of circumstances. They might as well hope to escape from their shadows.'[1] The very countrywomen, with their gold ornaments and broad-frilled lace caps, were too wealthy for our young doctors to consort with. Nobody knew them to be learned men, unless by recognition of the dazzling doctors' hats: they knew nothing of the mental treasures these poor tramps carried with them.

They followed on the road lined with brushwood, behind which a silvery line proclaims the Zuyder Zee, until the land breaks into hedges and falls into water-meadows, and meres replace the bogs. The land here

[1] Froude.

is more varied in its produce; there are standard fruit trees with currant bushes growing beneath them; and there are more silver-helmed peasant women about, and grander females with gold frontlets and engraved-plate head-bands and earrings; and here sound the sweet carillons of Sint Joriskerk at Amersfoort, where the young men rested to eat bread beside the canal rippling through the pretty town. On again across the sandhills, beyond which lies good ploughed land tilled with varied cereal crops, and the river Eem glides—for rivers never run in Holland—gently through a wood. The ground is slightly undulating here, so it is able to glide; otherwise it would become a 'mere' like the rest of the rivers. Beyond this again the land is sandy and in all stages of reclamation, with fir plantations and beech. Linnæus 'saw all that, and saw all that lay behind it—a miracle of human industry, two millenniums of human history.'[1] What a good description of the country hereabout is given in the name Watergrassmeer! The man was a genius who coined the word; the village here, with its pretty pleasure-houses set in bowers and ornamental waters, is an oasis among the sand-dunes!

Beyond the further marshy ground is a blaze among the colza. What a smoke! It is a damp reed-hut on fire. The travellers rested again at Weesp—a town fortified with grass terraces, and set as usual in a mere—and they arranged their travel-soiled dress and put on the gay green hats that they might enter Amsterdam with

[1] Froude.

becoming dignity. How fine is the view of Amsterdam seen beyond the watery landscape on entering the city by the Y, and how profuse is the distribution of gold jewellery among the well-to-do womenfolk! Alas for these poor young doctors, who can hardly muster a stiver between them, and who cannot even exercise their new profession for want of knowing Dutch! They must look out for learned patients who can converse fluently of their ailments in the Latin tongue. A poor prospect for Linnæus.

The pressing questions are, How can he work his passage home to Sweden? and How will his Elizabeth's papa receive him when he gets there? He faced both questions like a man. It seems Clæs Sohlberg had cash enough left to go home with, for we hear no more of him, nor of any difficulty in his finances. Probably a remittance awaited him at Amsterdam. Linnæus may not have been so good a man of business as Sohlberg, but then he had more to do and to see wherever he went, and sightseeing in towns involves fees. Perhaps, too, he was more unskilful in paying and giving away. Goldsmith, writing in 1759, 'Would you believe that in Sweden highway robberies are not so much as heard of? For my part, I have not seen in the whole country a gibbet nor a gallows!' Linnæus, used to his own truthful people, was possibly often taken in.

CHAPTER XI.

LEYDEN—THE FAT OF THE LAND.

*When lands are gone and money spent,
Then Learning is most excellent.*

CARL'S first business was to see if anything had turned up since his previous eight days' stay in Amsterdam, where he had tried to place the MS. of his 'Systema Naturæ' to advantage, and to present his letters of introduction to the rulers of the scientific world. He was anxious to make himself known to the leading Dutch naturalists before he returned home. Wearing his gay hat, he waited on the professor of botany, Dr. Burmann, with no immediate result; but, as Carlyle says, 'Hope *diminished* burns not the less brightly, but like a *star* of hope.'

He afterwards proceeded through Haarlem to Leyden, where he visited the botanical garden and Professor van Royen; but of all the persons Linnæus met with in Holland none paid him more attention than J. Fred Gronovius, doctor of medicine.[1] Carl having paid him a visit, Gronovius returned it, and saw the sketch of

[1] Diary.

his 'System of Nature' in MS., which so astonished him by its novelty, that he requested Linnæus's permission to get it printed at his own expense.

What a triumph for the youthful doctor! His talents were to be recognised at last. What mattered poverty or even hunger now? He was to be known as the author of the New System! He forgot that few people saw the necessity for a new system, or indeed for any system in botany at all.

The publication of the work was accordingly commenced—a matter of immense importance to his after fame, and really better than bread and Dutch cheese in the present. But an engraved work finely got up on fourteen folio pages [1] takes some time to prepare, and to distribute it profitably takes still longer.

Though Carl wanted so little, it seemed as if his life were always to be 'a progress from want to want, not from enjoyment to enjoyment'; that he must still cast about him to make something out of nothing, daily to twine his rope of sand. Still the greater, harder work, the chain of linked thought, was in process of production, and it was the first time he had tasted the exquisite cup of realising his dreams.

It was hard that he could not afford to remain in Holland till the birth of the great System, but he could not live without work, and paying work had yet to be found.

In this year, 1735, he published the first edition of

[1] Twelve folio pages, says Sir W. Jardine.

his 'Systema Naturæ,' consisting of eight large sheets, in the form of tables;[1] this edition is now a great bibliothecal curiosity. It contained a view of the animal, vegetable, and mineral kingdoms, and was the germ of that scheme of natural history which was a few years afterwards adopted throughout the world. 'In this way was the foundation laid of that system upon which almost all those of the present day are in many ways most intimately connected, and by which the arrangements of the older systematists were almost at once superseded.'[2]

By the advice of Gronovius,[3] Linnæus waited on Boerhaave in Leyden. Carl had particularly wished to see this eminent man, who was renowned throughout the world, so that a letter reached him from the Emperor of China, directed simply, 'To Boerhaave, the famous physician in Europe.' Not so very long after this it was Linnæus's own turn to meet with similar recognition.[4] Bjoernstahl saw at Therapia, in Turkey, a Greek walking in a field reading a book; the man (formerly first physician to the Pasha) told him it was by 'the great man in Europe.' It was Linnæus's 'System of Nature.'

Boerhaave,[5] through press of occupation, had great difficulty in granting audiences even to his friends. Peter the Great had waited several hours in an antechamber for an interview:[6] how could Linnæus, poor and a

---
[1] Carr.   [2] Sir W. Jardine.   [3] Diary.
[4] Carr.   [5] Ibid.
[6] This was many years before—in 1716, or perhaps in the Czar's earlier visit to Holland, in 1697.

stranger, hope for admittance? He awaited in anxiety the result of his application. 'He who had lived in Hamburg too high for his means, in Leyden had to live low.' He was wiser and therefore humbler now. His lesson in Hamburg had taught him that a *novus homo* must not be arrogant when he enters the society of the scientocracy, and that he must not run himself rashly against vested interests. Yet for all his poverty, Carl Linnæus seems to have lived in intimacy with the scientocrats of Leyden—Van Royen, Van Swieten, Lieberkuhn, Lawson, and Gronovius. Turton shrewdly says, 'Among the causes which contributed to enlarge the views and ripen the judgment of Linnæus may be reckoned the facility with which he made himself known and regarded by the most learned men of his time. Wherever he came he found a friend, and that friend generally of the first reputation in the sciences he studied.'

Days passed on, and Linnæus, having exhausted the sights of the 'Athens of the West,' was on the eve of leaving Leyden, when, on the eighth day after his first call upon Boerhaave, he was admitted to the physician's presence suddenly, and out of his turn, for several great people had been waiting longer than he. Learning was power here in Holland: whatever it may have been in his own country, here he was not without honour. It is true that in Holland, where one had only to 'invent a shovel and be a magistrate,' a new theory was certain to obtain respect, especially when it was about plants, which then meant tulips, which they adored. It seems that Gronovius had sent to Boerhaave a copy (or more likely the original MS.) of the 'System of Nature,' which made the great physician desire an interview with the young Swede. Boerhaave, then in his sixty-seventh year, received Carl with the greatest cordiality, and invited him to his country seat, a mile out of Leyden.[1] All elderly men relished the vigorous and far-reaching conversation of young Linnæus, and the freshness of his views, so well calculated to rouse their own flagging enthusiasm. Boerhaave,[2] one of the richest men in Leyden, was extremely plain and active, and a thorough Dutchman. His whole wardrobe consisted of two suits, which he wore till threadbare. His Dutch-built figure, standing in his old shoes, with his loose hair, and the large crab-stick which he carried, made him look like a common man. He was parsimonious, having been brought up in a frugal school; but he was very beneficent to the poor.[3]

He had a botanic garden, and collection of exotics, among which he pointed out one of the hawthorn family (*Cratægus aria*), and asked Carl if he knew that tree, which seemed to be remarkable in Holland. Linnæus said he knew it well in Sweden, where it was common.[4] As Boerhaave's garden was stocked with all kinds of trees that would bear the climate, Linnæus had an

---

[1] Boerhaave was born at Voorhout, two miles from Leyden. His father was minister there.

[2] Stoever.   [3] Carr.   [4] Diary.

opportunity of manifesting his skill in the science and history of botany. Boerhaave, observing this, advised him not to leave Holland immediately, as he had intended, but, on the contrary, to take up his abode there. Linnæus admitted he had not the means of remaining a single day longer. As Carl proposed passing through Amsterdam on his way home, Boerhaave, who wished to serve him, gave him a warm letter of recommendation to his pupil, Professor John Burmann, and desired him to present it with his compliments.

This altered the aspect of affairs: Carl's third visit to Professor Burmann was no failure. Next day Linnæus called to see the professor, who personally conducted him over his collection, asking him which particular plants he wished to inspect.

'Which of my plants do you wish to examine?' 'The greatest number, and even all of them,' said Carl, ' but I do not know which plants you possess.'[1]

The botanic garden of Amsterdam, formerly so celebrated, is now scarcely worth the notice of an English botanist. It is, however, neatly kept, and contains some good specimens. The Dutch in general seem still to retain that extravagant rage for buying rarities at an exorbitant price, for which they have long been famous.[2] A fine street in Amsterdam, leading to the botanical garden quarter, is named the Linnæus Street, and a mile beyond the Muiderport is the Linnæus

[1] Diary.   [2] Sir J. E. Smith.

Garden, a school of horticulture and forestry, where the glass-houses are kept in fair working condition, but not in apple-pie order for show.

To return to our friends, 'This is very rare,' said Burmann, pointing out a plant in his herbal. Linnæus asked for a single flower; he softened it in his mouth, examined it, and pronounced it to be a species of laurus. 'It is not a laurus,' said Burmann. 'But it is,' said Linnæus; 'it is the cinnamon tree.' 'It certainly is the cinnamon,' rejoined the other. Linnæus then convinced him that this tree was a species of laurus, and also corrected his classification of other plants.

Burmann was at this time preparing his 'Thesaurus Zeylanicus,' a great work on the plants of Ceylon,[1] and he was so charmed with Linnæus that he offered him a handsome apartment, with attendance and his table, if he would be his guest and help him with his book.[2] Linnæus availed himself of these advantages until the following year. Burmann had a fine collection of natural curiosities, and a well-chosen library. Carl took the opportunity of studying them to complete and publish his own 'Fundamenta Botanica,'[3] a small octavo volume of thirty-six pages, in the form of aphorisms, which contains the very essence of botanical philosophy.[4] Linnæus says he amused himself with looking over

---

[1] The *Flora* of Ceylon, though rich, has scarcely proved so voluminous as was expected; yet it comprises 3,000 plants. Ireland, a somewhat larger island, has only 800 kinds of plants.

[2] Diary. [3] Signed C. Linn., Stipend, Wredian. [4] Carr.

Burmann's works on the plants of Ceylon, and frequently visiting the botanic garden.

Mr. George Clifford,[1] J.U.D., burgomaster, banker, and one of the directors of the Dutch East India Company, was at this time the most enterprising botanist and horticulturist in Europe. He had a fine country seat and garden at Hartecamp, near Haarlem.

He was out of health, and applied to Boerhaave for advice. The doctor recommended Linnæus to him as one capable of looking after his health, and who would also be able to arrange his fine collection of foreign plants and form his garden, which cost the banker 12,000 florins annually, and was his hobby and his pride. All Dutchmen love their gardens, but Clifford was no common tulip-fancier, but an ambitious man of scientific aims—one of the men, Motley's republican Dutchmen, the makers of Holland, who made their small country a leader in European history. Hartecamp was no ordinary *lusthaus*, as Boerhaave well knew; he also rightly judged that the eager enthusiastic young Swede was no mere classifier of plants in a herbal, but one who would tend and keep in order a paradise. We are not to suppose that, for all Carl's science, ' a primrose by the river's brim,' a 'monopetalous hypogynous Pentandria monogynia' was to him, and it was nothing more: on the contrary, he was first and above all things a florist. He kept the dried flowers in his herbal and

---

[1] Stoever and his copyists, following the German pronunciation, write the name Cliffort. Dutch books spell it Clifford.

wrote a descriptive epitaph upon them, as we embalm the memory of our friends and adorn their graves.

> And 'tis and ever was my wish and way
> To let all flowers live freely and all die,
> Whene'er their genius bids their souls depart,
> Among their kindred in their native place.
> I never pluck the rose; the violet's head
> Hath shaken with my breath upon its bank,
> And not reproacht me; the ever-sacred cup
> Of the pure lily hath between my hands
> Felt safe, unsoiled, nor lost one grain of gold.[1]

Clifford visited Linnæus at Burmann's, and invited them both to come to Hartecamp and see his hothouses and his Cape plants. This was a real pleasure to Linnæus, as after the dead levels about Amsterdam, the more undulating country round Hartecamp afforded an enjoyable change of scene. He roamed through the gardens with a boy's delight, and examined the hothouse treasures, describing those that were known, and speculating on those that were new; while many a truth fell from his lips, 'contained within the concise limits of a passing jest,' in sportive vein, wreathed in dimpling laughter, showing in all simplicity his enjoyment of the holiday. He was one of the most loveable of young men. Clifford was equally delighted with him and his agreeable way of imparting knowlege—which argues a familiar knowledge of Latin on Clifford's part.

Burmann took up in the library the second part of Sir H. Sloane's 'History of Jamaica.' 'I have two

[1] *Fæsulan Idyl*, W. S. LANDOR.

copies of it,' said the banker, laughing, 'and you shall have this if you will give me Linnæus in exchange.' The proposal thus made, apparently in jest, soon led to a serious one, and Clifford invited Linnæus to live with him as his physician and botanist, and offered him a salary of 1,000 florins per annum.[1]

Linnæus was dazzled and captivated by this new experience. Never had he met with a sort of life so tucked in with velvet curtains, such sumptuous appearances of equipage and well-laid table, and everything so rich and bankery: but these things alone would never have tempted him had there not been ample liberty and the garden at command, and unlimited powers given him to use both to the best advantage. He could here cultivate science without restriction. He was truly happy. Hear himself: 'Thus Linnæus moved to Clifford's, where he lived like a prince ; had one of the finest gardens in the world under his inspection ; with commission to procure all the plants that were wanting in the garden, and such books as were not to be found in the library; and of course enjoyed all the advantages he could wish for in his botanical labours, to which he devoted himself day and night.' His energy had a tremendous impulse now that he was settled and at leisure. He was an excellent companion too. He 'had an immense fund of articulate gaiety in his composition, beautiful light humour,' never flying off into folly, 'yet

---

[1] Turton says 800 florins a year; Fée says 1,000 florins. We must take that sum which is nearest a ducat a day.

full of tacit fires which spontaneously illuminated all his best hours.' This, which in his wife was such a charm to even the serious Carlyle, is a good description of the gay gleams which Carl Linnæus flung over a life which other scientific men contrived to render dry as dust. He could throw himself into wildest spirits in off-work hours. He would imitate the contortions, grimaces, and incantations of the Laplanders until his audience thought his acting equal to his science. Clifford felt he could never do enough for a youth who regilt life for him, wreathing it with flowers the while, and bringing back all the best aspirations of his younger days. The golden head brought back summer to the rich man, whose hair was already just flecked with snow, and showed him he still could enjoy 'more, indeed, than at first when unconscious, the life of a boy.' He loved Carl like a son, and gave him (what Carl most valued) his duplicate dried plants.[1] Does this seem a bathos? It is really none. Now was Carl's time to bring forward his 'Critica Botanica,'[2] his 'Genera Plantarum,' and to commence a fine folio volume called 'Hortus Cliffortianus'—a complete catalogue, splendidly illustrated, of the garden at Hartecamp.

Tulips do not seem to have been of much account at Hartecamp, though I dare say the 'Admiral Enkhuizen,' valued at 4,000 florins, the 5,000 florin 'Admiral Liefkenshoek,' and the famous 'Semper Augustus,' costing 13,000 florins, displayed their splendour in the conser-

---

[1] Diary.   [2] In one vol. 8vo. This book is very rare.

vatory; but the mania of a century before, 1636, when tulip-roots passed from hand to hand like bank notes, could never be revived.

The Jacobea lily (*Amaryllis formosissima*) now first blossoming in Europe, was a gem of the cool greenhouse and 'nature herself favoured Linnæus in causing through his diligence and care the fine plantain tree,[1] also to bloom in Holland for the first time, which was looked upon through the whole country as a wonder.[2] Even Boerhaave himself came to Hartecamp to get a demonstration of this *musa*; described for posterity in the treatise that was afterwards published under the name of *Musa Cliffortiana*, whereby every gardener has been enabled to bring forward its flowers.'[3] Linnæus here acquired great practical knowledge of plants, including palms, which he called the princes of vegetation.[4] He visited every month the gardens of Amsterdam, Utrecht, and Leyden, but every day that of Hartecamp.[5]

The situation of Hartecamp is the pleasantest in Holland; it has the sandhills of the North Sea for its horizon on the west, from which quarter the breeze blows during the greater part of the year and bends the trees landward. The young Swede rejoiced in the sea

[1] *Musa paradisaica*.   [2] Stoever.
[3] This tree flowers at Kew Gardens Oct. to Dec. The fruit begins to set in April.
[4] 'Man *dwells naturally* within the tropics, and lives on the fruit of the Palm tree; he *exists* in other parts of the world, and there makes shift to feed on corn and flesh.'—LINNÆUS on Palms. 'Honour the date-tree, for she is your mother.'—MAHOMET'S commandment.   [5] Diary.

wind, from which his garden, however, was protected by the lines of overarching elm-trees, and the sand-dunes piled high beyond the water-meadows stocked with black-and-white cattle. Carl, though he had a carriage and four horses at his command, I dare say went oftenest to and fro between Hartecamp and Leyden by the barge, in the canals narrowed by the rapidly growing sedges, water-flags, and lilies; where the labour is ever going on of dredging black mud into boats, then filling it into trough-shaped carts, or else plastering it upon the banks—the canals covered with white water-lilies expanding their unsullied flowers to the morning sun, and intermixed with the yellow-fringed water-lily, which is very uncommon in England. The silence that accompanies the Dutch (canal-boat) mode of travelling, so different from the grating of a turnpike road, increases in no small degree the pleasure of a journey.[1]

One can now go to Hartecamp from Leyden by train to Vogelenzang, and then inquire the way to Bennebrock. It is best to follow the peasant girls who get out at the station; they are most likely going to pass the Hartecamp, as it lies on the main road to Haarlem. From what I had read I expected to find Hartecamp a pair of iron gates, a swamp, and perhaps an avenue; but it is by no means the howling wilderness that writers represent it. The ground at Vogelenzang rises in pleasing undulations, chiefly of reclaimed sandhills clothed with fir trees, which wave refreshing scent beneath

[1] Smith.

the silvery cloud-banks rolled up massively to show the blue; the rich low ground all laid out in bulb gardens. German goes next to no way in Holland, and no one understood that I wanted to see the place where Linnæus once lived—how should they? Luckily, a young gentleman came up who spoke a little German. He showed me the way to Hartecamp (he lived in the next villa but one himself), and he spoke to the gardener for me. He pointed out the name 'Te Hartecamp' on the gates and told me this was the actual and nearly unaltered house of Clifford. The plant-house or architectural conservatory—something like the one they house orange trees in at Kew—was also Clifford's, and the fine vine with the thick stem was here in Linnæus's time. A row of fine bushy and aged Portugal laurels grows in front of the architectural conservatory. There are grand timber-like oleanders, which look as if they might have seen the great botanist, and ancient orange trees in tubs, and purple clematis twines about the pilasters. We hear it is an easy walk to Haarlem, which makes us independent of trains and able to enjoy the lovely garden.

The guide-books are wrong in saying it is a waste place or wilderness; though the glory of Hartecamp perished with Clifford, it is a fine garden still, with alleys and avenues in all directions, and winding sea-shell-sanded paths by the ornamental water, enlivened by swans and crossed by a fanciful bridge. Bædeker says, 'the beautiful gardens attached to the house have long since disappeared.' He must be either extremely curious in gardens or else he visited Hartecamp through an agent. It is a delightful place, not being on such a dead level as the rest of Holland. The house is shut up during half the year, while its owner lives at the Hague.

In the pleasant deep bay-window of the central ground-floor room, at the back of the house, or on the great balcony above, Linnæus often worked, and looked out upon the lawns and lakelet encircled with great purple and copper beeches, and variegated horse-chestnuts, which have white leaf-masses near the trunk and thick stems, though the rest is green, being able to suck in the sunshine better from being exposed. The windows command views through glades right away to the dykes and dunes.

The flower-border verges are left fringed with wild plants of all sorts, spreading into and embroidering the lush spring lawns. How we enjoy the delicious countrified look of all the plants and trees (we hail from London and Leyden), and having it all to ourselves like this!—for the young gentleman told the gardener to leave us alone to sketch. This civilised verdure—if one may so express it—is enchanting. From the avenue where I sit I see a brilliantly coloured vista of foliage beyond some emerald green elms, and one tree all over white blossoms, an oak with golden-tufted buds, and then an amber-coloured tree, and beyond again a clump of crimson beeches. The trees all round are grouped with great taste—some of them knew Linnæus; the borders are botanically arranged, as if the Linnæan traditions still held sway. The path leading from the front of the house to the conservatory has a border set amphitheatre-wise with lines of rare plants in sunken pots, sheltered by a pine grove. The foregrounds are massed with clumps of *gunnera scabra* and lady fern. The air is full of sounds of birds, the swans float by and halt, and lose themselves again in the thickets. The lawn in front of the house is planted like a park with various trees, well arranged with an eye to colour. The road, itself invisible, crosses the view, so that one can see the picturesque passers-by, who animate the scene at just the distance an artist would set them in his picture; beyond the road the park sweeps upward to a belvedere, high-raised (for Holland, that is) at the end of the vista. An immense oval bed of roses is spread just before the house.

'One should think that the proprietor of all this must be happy.' 'Nay, sir,' said Johnson. 'All this excludes but one evil—poverty.' So Clifford felt till Linnæus came. Carl enjoyed it all without the cares attaching to ownership. I like to picture him with this pleasant background about him. The front door, as usual in houses of that age, is in the centre of the house; the steps are flanked with large ornamentally painted tubs of palms, aloes, and masses of New Zealand flax. The inlaid marble pavement before the house is so deeply buried in sand, the driftings of the last few months, that it shows how important a factor the wind is in the making of Holland. It has a far quicker action than Darwin's earth-worm works. This pavement would be buried and grass-grown in a year.

A deer, the Harte of the Hartecamp, points the vane above the clock on the top of the house. The enamelled white furniture of the villa is partially the same as in Clifford's time. It is about four miles to Haarlem, and we proposed to walk there. Seeing a baker's cartlet drive up to the house, I rushed to buy buns and fancy bread for luncheon, but the housekeeper, wife of the gardener, waved a teacup at us, beckoning us into the basement, which I had supposed to be cellars. Here was a range of low-roofed but most comfortable kitchens, unaltered since Linnæus lived here; the brick-paved floors, skirted with white tiles, had raised wooden movable floors laid on the bricks, and at the doors large mats of thick basketwork or fine hurdlework: and many hints and contrivances for comfort, showing how they successfully resist the subsoil dampness of even humid Holland; and showing how comfortable daily life was in Holland, over a century ago, when we had far fewer of the minor luxuries. The good woman gave us bowls of coffee and milk, and then unlocked a side gate beyond the wood to show us the ' kooter way to loopen na Haarlem.'[1]

It is a pleasant walk through the pretty woodland, on a good road lined with country houses and closely paved with small bricks, nice to 'loop' on, as is usual in the main country roads hereabout. These roads must have been made at a frightful expense, but they

---

[1] Not pure Dutch, I fancy, but as it sounded to us.

were worth making. The ground about here is broken and uneven, with fir trees on the hillocks, all of which gives it a picturesqueness Holland generally lacks. The next house to Hartecamp, which is now one of a series of villas (!), has a lodge with statues and other ornaments, in the questionable taste of the eighteenth century. Near this is an obelisk to the memory of Count Floris van Zoon van Holland, and others who fought and fell with him. The way still passes between gay villa gardens with a woodland background. The character of the country is quite different from what it is about Leyden, though it seems to be only reclaimed dunes. This road, lined here with florists' bulb gardens, enters Haarlem close by a tall modern church, a turnpike, and the Flora Park, whence the tramway goes directly through the town to the railway-station. In Linnæus's time all the Haarlem world was talking of the great organ erected in the cathedral in 1735, this very date, at the town's expense. Linnæus never mentions it, but, as we know, he was not fond of music.

In the Teyler library and museum is the original portrait of Linnæus in his Lapland dress, which was painted from life at Clifford's. Several copies were executed, and a print [1] of it is in the Linnæan Society's rooms in London. It represents him with boots of reindeer-skin; about his body is a girdle, from which is suspended a Laplander's drum, a needle to make nets, a straw snuff-box, a cartridge-box and a knife, a grey

[1] Not a copy, as has been erroneously asserted.

(or brown) round hat, and brown wig. He wears Laplander's gloves. This portrait shows a wart on the right cheek. It is altogether the most pleasing portrait of him that we have, representing a good-looking brown-eyed young man, of serious but intelligent expression, aged twenty-eight. The lively colours of his garments are a blue collar lined with red, and a yellow worked yoke below the collar, a blue pouch, red watch-bag with yellow top, brown dress, green and yellow scalloped leather case for tools or collections, for which purpose he doubtless utilised also the Laplander's drum. He holds the pink flower, which had just been published under the name of *Linnæa borealis* by his friend Gronovius. An engraving of this plant is given in the twelfth plate of the 'Flora Lapponica,' which Linnæus had succeeded in getting printed by means of a society at Amsterdam of which Burmann was a member, and which Linnæus had often visited, the society offering to advance the twelve plates,[1] which are interleaved with verses as mottoes. Some of these are in Swedish, but they are chiefly from Ovid and other Latin poets. The andromedas figure in the first plate. The first page has some gushingly complimentary verses from Brouwallius, dated from Fahlun in Suecia, November 24, 1736, to his 'peerless friend Carolus Linnæus, Med. Doc.' The frontispiece to Smith's edition of the book is a Laplandish willow-pattern-plate sort of landscape: some precipices, like ruined steeples set in substantial clouds,

[1] Diary.

are nearly upset by the solid rays of the midnight sun; Laplanders are rushing about in wildest action with their canoes, tents, and other attributes; a reindeer *couchant* in the foreground supports a Laplander dining off his drum. The *Linnæa borealis*, a dried specimen, is in the right-hand corner. Linnæus divides Lapland into two regions—Alpes Lapponicæ and Desertum Lapponicum. Preparing this work had been Carl's recreation amidst his severer studies. Oppressed at times by the weight of luxury around him, and by the heavy climate of Holland, he revelled in the recollection of the Lapland Alps—for the memory of hardships had now become a pleasure. Except in the immediate society of Clifford and the scientific men who gathered round his table, Carl was very much thrown inwards upon himself. One wonders he had not shrunk narrower, thus dwelling in a small coterie of persons of one turn of mind exclusively, among whom he early took a leading part; and he would have done so had he not taken the whole range of natural history for his province, and held fast his idea of benefiting his own Sweden by his researches. 'Suddenly cast,' as Gibbon phrases it, ' on a foreign land, he found himself deprived of the use of speech and hearing; incapable, not only of enjoying the pleasures of conversation, but even of asking or answering a question in the common intercourse of life'; for, as of old, Linnæus, the inventor of words, never could learn words for words' sake. Like our Johnson, he only knew one Dutch word—*roes-knopies*, rosebuds. ' And that is Swedish too—*roes*, rose, *knopie*, knob.' When the rich banker's gardens became, as they sometimes were, the playground of a brilliant circle of fashionables from the Hague and Amsterdam: when the lawns and groves were crowded with modish folks with bright complexions, powder and patches, smiles, toques and turbans, tall ample-ribboned hats, trains and hooped petticoats, and all the paraphernalia of a breakfast party in the afternoon: the interesting but dumb young Swede at first shunned the band of youth and wit, and mingled with the fusty celebrities exclusively.

Although 'endowed by art or nature with those happy gifts of confidence and address which unlock every door and every bosom,' and solid learning besides, to give these airy graces weight, what could these things avail him outside the learned and masculine circle? Clifford enjoyed Carl's society intensely, and elderly men admired him. 'His gifts were just what Holland needed; here he was brilliantly successful.' Young men envied him from a distance, but women held him in too much awe. He possessed 'that flexibility of manner and readiness of gentle repartee' which would have made him delightful to young women, but that his talk was all in Latin. What a pity! Otherwise he could have talked quite as much nonsense as other people. What avails even a firework of wit if it is all in Swedish or Latin? But they did not know that the young mute with the bright eyes,[1] expressive

---

[1] 'His eyes, of all the eyes I ever saw, were the most beautiful,

countenance, and splendid reputation was engaged to a young lady living near the Arctic Circle. They might have won him from dry botany, but not from 'the fair flower of Falun.' He was as polished and graceful as the best of their adorers, even in that time, when a French and finished manner was accounted the acme of everything.

> I love to see in all their fitting places
> The bows, the forms, and all you call grimaces.
> I heartily could wish we'd kept some more of them,
> However much we talk about the bore of them.
> Fact is, your awkward parvenus are shy at it,
> Afraid to look like waiters if they try at it.

But after a while Carl relished his leafy silences being broken by music, his tranquil lilies splashed by yawl and gondola amid the glancing water. He eked out his words of broken Dutch with frolic grace as he threw off the dominie for the time, and showed he too could laugh and enjoy youth and life among gladsome things, as he and Bartsch, his friend, the only other youth among the *savans*, led the way among the glades and groves, with a lively following of beings all frivolity and fun. Yet all this while he carried next his heart his little pocket-book with his name and Elizabeth's, mysteriously written so that none else could read them.[1]

says Fabricius, speaking of him at fifty. What must they have been now at twenty-eight?

[1] The little almanack he used in Holland, containing his name and his love's name inverted and intertwined, is now bound in crimson velvet and prized as a treasure by the Linnæan Society.

These diversions never caused a break in his work: they only added to his difficulty in finding time. 'Creative genius is not a passive quality that can be laid aside or taken up as it suits the convenience of the possessor.'[1]

One day, while walking in the streets of Leyden, passing round the *hoek*[2] by the 'informatory,' as they translate a school, Carl unexpectedly met his own loved friend, his second self, Artedi, who had just come from England; and oh, what an outpouring in the dear old mother-tongue! So much to hear and tell; such struggles and successes, and on Artedi's side such continual disappointment. He told Linnæus 'he had spent all his money in London, and he was in want of more to purchase clothes and books, and also for the purpose of obtaining his degree and returning home, and he knew no means of raising it.'[3] Poor Artedi, with his golden dreams vanished! Not only alchemy had failed, as Linnæus had foretold it would, but learning too, though everyone had prophesied it wouldn't. In the phraseology of those days, he saw Linnæus, who had climbed the steps of the temple of fame, while he stood below on the muddy level of adversity. The prosecution of his studies had reduced him to beggary—life cost so dear in England. Could Carl put him in the way of earning any money?

'Linnæus comforted him with the assurance that as he was not now under the confined circumstances

---

[1] B. R. Haydon. [2] Corner. [3] Diary.

and the persecutions to which he was exposed at Upsala, he would take care that his friend should be assisted.'[1] He quickly cast about him for the means, first and foremost ordering out the coach-and-four. He thought of Burmann's 'Thesaurus,' which had been the beginning of his own prosperity. But no: another man had also a 'Thesaurus'—Thesauri were the fashion.[2] 'Albertus Seba, a German apothecary at Amsterdam, had a short time before requested Linnæus to assist him in completing the third volume of his 'Thesaurus'; but, being then employed at Clifford's, Linnæus could not accept this offer; and besides, this third volume intended to be printed related to fishes, which Linnæus liked least of all the branches of zoology.' Linnæus went to Seba with Artedi, with, I dare say, no little complacency at having a coach-and-four at his command; a troublesome equipage for going round the *hoeks*, but well calculated to assist the views of his friend—as doors open wide to admit a coach-and-four. 'He recommended Artedi to Seba as the first man in ichthyology. The work was accordingly put in Artedi's hands, with the promise of a handsome recompense.'[3]

Carolus found it is so pleasant to be called Carl again in the old familiar tongue, that he often went to see his friend, to chat with him of old happy miseries, which talk revived yet more his longing to go home— to Sweden, home and beauty. While Artedi was painfully contrasting their lots Linnæus was beginning to

[1] Diary.   [2] Ibid.   [3] Ibid.

feel satiated with luxury, and loved best to talk of other things in life than wealth can buy. 'To live, in the true sense of the word, is to feel, love, desire, admire, and not to breakfast, dine, sleep, and yawn.' For the first time Linnæus cared to study the phenomena of human emotion. He longed for liberty and home, and did not feel he was so greatly to be envied. To himself he seemed like a lap-dog on a velvet cushion, who would prefer straw with its wholesome friction, and Artedi, though he now lived comfortably at Amsterdam and liked his work, yet felt it was not for his own fame, and looked forward likewise to his own return to Sweden. Artedi was out of heart about himself and doubtful of his own powers. His reception in England had been freezing. He felt 'remote, unfriended, melancholy, slow.' He had not the animal spirits of Linnæus; and, as Carlyle says, 'there is no fairy gift like this for helping a man to fight his way.' To his countryman's chagrin, he kept without the pale of the gay circle which welcomed the brilliant paradoxes of his exuberant and irresistible friend, and he seldom visited Hartecamp.

'No sooner,' says Linnæus in his diary, 'had I finished my "Fundamenta Botanica" than I hastened to communicate them to Artedi. He showed me on his part the work which had been the result of several years' study—his "Philosophia Ichthyologica," and other MSS.' On these Artedi had built his hopes, and these he could not bring to light for lack of the gold he had failed alchemically to make. He began now to think of himself rather than his aims—

> To dare let down
> My strung, so high-strung brain, to dare unnerve
> My harassed o'ertasked frame, to know my place,
> My portion, my reward, even my failure,
> Assigned, made sure for ever! To lose myself
> Among the common creatures of the world,
> To draw some gain for having been a man.
>   BROWNING'S *Paracelsus*.

'I was delighted with his familiar converse,' says Linnæus, 'yet meanwhile, overwhelmed with business, I grew impatient at his detaining me too long. Alas, had I known that this was the last visit, the last words of my friend, how fain would I have tarried to prolong his existence!' It was September 25.[1] Artedi had at length so far completed his undertaking for Seba that only six fishes remained to be described. This evening he was in company at Seba's, and on leaving Seba to return to his own home he fell into a canal and was drowned. The night dark, unknown the way, he came to the brink of a canal not enclosed by rails. His calls for help unheard, next day his body was found.[2] As soon as Linnæus heard of this he went to Amsterdam to see what could be done to honour the name of his poor dead friend and save the ichthyological MSS., to which he was heir according to the

---

[1] Diary.
[2] Preface to Artedi's *Philosophia Ichthyologica*, edited by Linnæus.

will Artedi had executed before both the friends left Upsala.

'The landlord, however, having made out a bill to the amount of more than 200 guilders, Linnæus went to Seba and tried to prevail on him to redeem the MSS.; but the latter would give only fifty guilders towards the burial of Artedi. Linnæus then persuaded Clifford to advance the money'; and himself afterwards raised the best monument to his friend's memory by finishing and publishing Artedi's work on ichthyology, with a pathetic account of his drowning in a foreign country in the preface.

> Ill-fated youth! on whose unclouded brow
> Hope faithless gleamed, to lure thee to thy doom;
> And made thy various busy race below
> But a more speedy transit to the tomb!
>
> . . . . .
>
> And art thou gone? Are all thy virtues dead?
> Oh, no! for Heaven's eternal justice reigns!
> Thy buds of Hope, though plucked, shall never fade;
> Their fruit shall ripen in celestial plains![1]

The death of his bosom-friend was a bitter loss to Linnæus, who now began to feel the cruelty and silence of exile.

[1] Translated from a poem on the death of Pehr Artedi.

CHAPTER XII.

A VISIT TO ENGLAND.

Wearily stretches the land to the surge, and the surge to the cloud-
  land;
Wearily onward I ride, watching the water alone.
Not as of old, like Homeric Achilles, κύδεϊ γαίων,
Joyous knight-errant of God, thirsting for labour and strife,
No more on magical steed borne free through the regions of ether,
But, like the hack which I ride, selling my sinew for gold.
Fruit-bearing autumn is gone; let the sad quiet winter hang o'er
  me—
What were the spring to a soul laden with sorrow and shame?
Blossoms would fret me with beauty; my heart has no time to be-
  praise them;
Grey rock, bough, surge, cloud, waken no yearning within.
Sing not, thou skylark above! Even angels passed hushed by the
  weeper.
Scream on, ye sea-fowl! my heart echoes your desolate cry.
Sweep the dry sand on, thou wild wind, to drift o'er the shell and
  the seaweed:
Seaweed and shell, like my dreams, swept down the pitiless tide.
                                        Elegiacs, KINGSLEY.

LINNÆUS's longing for Sweden and the Lapland Alps was smothered for a while in a change of scene that occurred to him in the spring of 1736, turning his thoughts away from the North. Clifford, ever kind to him, saw his depression, and thought change of air would be beneficial to his favourite. Accordingly, his employment at Hartecamp was varied by a journey to England, at Clifford's expense, to see the nursery-grounds of London and Oxford, and the North American plants cultivated in both places. This scheme promised him an interesting comparison of plants growing in the same latitude and the same hard climate as Sweden, as well as the sight of some newly-imported specimens of a vastly richer Flora in the juxta-tropical zone on the other side of the Atlantic.

Sir Hans Sloane was at the head of natural history in England, and to him Linnæus carried a warm recommendation in a letter of introduction from Boerhaave, couched in flattering terms—an unusual thing as coming from Boerhaave. It was written in Latin, in this style: 'Linnæus, who will deliver to you this letter, is alone worthy of seeing you and of being seen by you. They who witness your meeting will behold two men whom the world can scarcely equal.' This elegant letter may still be seen, by anyone who takes a good deal of trouble about it, in the British Museum.

Carrying the precious letter in his pocket, Linnæus embarked at Rotterdam for Harwich. The run down to Rotterdam shows some ultra-Dutch landscape scenery, with bright gleams of Cuyp-like sunshine upon it. In Holland one always thinks of the painters; yet perhaps as pretty a scene as any, and as truly Dutch, although no old master has translated it, is the view on the Boompjes, looking across the moon-lighted river to the willowy bank on the other side; the whole seen through a veil formed of the rigging of the shipping, mingled with the darker branches of the trees. Carl thought to cross over to London in one day, and expected to be away eight days altogether; but he had to wait for a vessel, and, owing to rough weather, he only reached Harwich in eight days after leaving Rotterdam. Linnæus was never sea-sick, which accounts for his being able to talk so much about the heathen gods— unless, indeed, it is Stoever who here shows off his mythological knowledge on this appropriate occasion. From Harwich Carl went by land to London—by coach probably; or did he, being in funds, enjoy the learned luxury of a post-chaise? The coach-road runs by the river Stour to Colchester, by Tiptree Heath—Tiptree of Mechi fame (how the experimental farming there would have interested our Swede!)—by Witham and Chelmsford, by Ingatestone, Brentwood, Romford, and Stratford. Entering London by way of Bow, and passing by that 'strange anarchy of a place, the Stock Exchange'— Carlyle's Domdaniel—he reached Charing Cross, then, as now, the flood-tide of human existence. 'The London street tumult has become a kind of marching music to me,' says Carlyle. Linnæus spoke of London in the only language he knew besides Swedish, which counts for nothing out of Sweden, as 'Punctum saliens in vitello orbis.'

With neatly-arranged dress—bloom colour, no doubt —ruffles, and dress sword, and the pretty letter in his bosom, Carl soon found his way westward to Chelsea.

Notwithstanding his stylish appearance Sir Hans Sloane received him none too warmly. He had been king of natural history too long to care about welcoming a possible successor, especially one who would try to upset all his arrangements. Fascinating men 'are apt to disturb the world.'

Sir Hans was getting too old to enter into new theories, and Linnæus's bold attempts (for he had heard of him otherwise than through Boerhaave) to introduce a new system of nomenclature excited in him more jealousy than admiration.[1]

Cake and wine of course were offered—such was then the fashion for morning visits; but Latin, as it is spoken, is hard to be understood between speakers using a different pronunciation. It is easy enough to those already used to Italian; but although the Latin language is much easier for scientific intercommunication than French or German, talk is still uneasy to English Latinists. There is nothing more absurd than the modern crusade against Latin or Greek by people who deem science the only useful education. Considering that with every science we have to learn its terminology, it is absurd to think we can do better than learn those languages, which are the alphabet of all science. Who can even read a book on anatomy unless he has studied Greek and Latin? The same with chemistry or mineralogy: every third word is in an unknown tongue. Any foreigner who knows Latin can read an English

[1] Stoever.

book on science easier than an Englishman who knows no Latin. Every French and German book worth reading is translated, but into a language founded altogether on Greek and Latin, and *only the words of one syllable are changed*. Never have Latin and Greek been more useful than now, if not absolutely essential: the first chapter of any scientific book will prove this. Latin and Greek represent less the language of the classics than the language of science. And for this we have to be grateful that the great nomenclator of science was a Swede, whose language does not pass current out of Sweden. Had so great a man been either English, French, or German, he would have tried to impose his own language on a rebellious world, and science would have had no neutral ground. A confusion of tongues would have been an infinite loss to science. What term would have had exactly the same shade of meaning in another tongue? There was never a time when Greek and Latin were more needful to be learnt than now; not as grammatical exercises, but for the words themselves. Without them one must be dumb or childish, asking the meaning of every other word. Without them one cannot add a word to the scientific vocabulary. The language of learning is studied not so much to read ancient lore as to understand and create modern science. The dead languages were never more alive than now—since Linnæus began the resurrection of the dead languages.

'Of all professions, the medical profession is most scientific, but if you read a modern medical book you find a hundred new terms, Greek all of them, all of them incomprehensible to Anglo-Saxon readers. Why do they warn us off the "dead languages," as they call them, and then wrap up all their wisdom in Hellenic words?'[1]

Although Carl's visit to Sir Hans Sloane was a failure, there was another person in Chelsea to whom he also carried an introduction. This was Philip Miller, the since celebrated gardener to the Society of Apothecaries. Fée gives this account of the interview (direct from Linnæus): 'When I paid Philip Miller a visit, the principal object of my journey, he showed me the garden at Chelsea, and named me the plants in the nomenclature then in use, as for example; "*Symphytum consolida major, flore luteo.*" I held my tongue, which made him declare next day, "That botanist of Clifford's does not know a single plant." I heard this, and said to him just as he was going to use the same names: "Do not call these plants thus; we have shorter and surer names—we call them so-and-so." Then he was angry, and looked cross. I wished to have some plants for Clifford's garden, but when I came back to Miller's he was in London. He returned in the evening. His ill-humour had passed off. He promised to give me all I asked for. He kept his word, and I left for Oxford after having sent a fine parcel to Clifford.'

Of this Chelsea garden Hare says,[2] 'The Botanic

---

[1] Bishop of Oxford on Language, Feb. 11, 1886.
[2] In his *Walks in London*.

Garden, facing the river, is the oldest garden of the kind in existence in England, Gerard's garden in Holborn [1] and Tradescant's garden at Lambeth having perished.' The Chelsea ground was leased to the Apothecaries' Company (who still possess it) by Lord Cheyne in 1673, and was finally made over to them by Sir Hans Sloane in 1722. Evelyn used to walk in 'the Apothecaries' garden of simples at Chelsea,' and admire ' besides many rare annuals, the tree bearing jesuits' bark, which has done such wonders in quartan agues.' A statue of Sir Hans Sloane was erected here in 1733. Near it is one of the picturesque cedars planted in 1683; its companion was blown down in 1845.' They were conspicuous objects.

Our Carl found a great deal to talk about when actually in the garden with Miller, concerning the English Flora. He found a range of plants new to him in those that grow upon the chalky soil of England. Sweden is almost destitute of chalk, and the parts of the Low Countries that he knew have no more. Carl had never yet seen the line of white cliffs by Ostend. It is vastly different talking in the open air with an energetic young man, to sitting ceremoniously in a room with an elderly gentleman who is rather bored than otherwise by having to entertain you. The difficulty with the language was perhaps greater; but the language of signs, of play of feature, and, above all, of sympathy, goes farther than neatly-turned Latin.

---

[1] Gerard, called the Father of English herbalists.

Miller knew, besides, plenty of gardeners' Latin, and that first day they do not seem to have squabbled.

They were proud of their hothouse at Chelsea, though it was no longer the unusual thing that it was when Evelyn spoke of it as so 'very ingenious, that the subterranean heat, conveyed by a stove under the conservatory, all vaulted with brick, so as he has the doores and windowes open in the hardest frost, secluding only the snow.' They were trying ineffectually to grow the *Ricotia Ægyptiaca*: Linnæus recommended them to mix Nile mud in the pot. Linnæus enjoyed his visit so well that he repeated it often—not, I suppose, in his bloom-coloured coat, but in thrifty work-a-day dress; though I dare say he donned the bloom-colour when he went in the evening to Ranelagh Gardens close by, deemed by Dr. Johnson himself ' a place of innocent recreation.' It must have been pretty much like the ' Healtheries ' and succeeding amusements have been in our time, or like the evening *fêtes* at the Botanical Gardens: not so low or lively as Cremorne, as they only danced the minuet at Ranelagh.

Boswell, comparing it with the Pantheon, of which we read so much in ' Evelina,' says, ' The first view of it ' [the Pantheon] ' did not strike us so much as Ranelagh, of which he ' [Johnson] ' said the *coup d'œil* was the finest thing he had ever seen. The truth is, Ranelagh is of a more beautiful form ; more of it, or rather, indeed, the whole rotunda, appears at once, and it is better lighted.'

Johnson expressed himself 'a great friend to public amusements, for they keep people from vice.' Doubtless Linnæus, who was of a lively social turn, relished these things, and most likely would have had Dr. Johnson pointed out to him. Though we never read that they met, they might well have done so, here or at Lady Ann Monson's house, and they could have talked Latin fluently together. The Hunterian Oration, which was then always delivered in Latin, was a subject of interest for Linnæus. Johnson's name was already immortal, but, although John Hunter was even then a distinguished representative of British surgery, the world did not yet know that both Hunter and Linnæus were as great as the lexicographer himself.

Lady Ann Monson, herself a lady of talent, and a botanist of no mean order, was very kind and attentive to Linnæus, who named a beautiful plant *Monsonia* in her honour.

There was plenty of gaiety going on in London; for May has always been the London season, and this was May. Linnæus saw Garrick act, and he saw the lions at Exeter Change, and on Sunday he went to the Foundling Chapel and heard the *Te Deum, Jubilate*, and an anthem (on occasion of the charity sermon) composed by George Frederick Handel, Esq., and performed under his direction, where, because of the pressure of the crowd, 'The gentlemen are desired to come without swords and the ladies without hoops.'

Linnæus seems greatly to have enjoyed Chelsea; and

that the officials here appreciated him is shown by the Chelsea garden being the first in England arranged after the Linnæan system. 'Miller allowed me to gather in his Chelsea garden, and gave me, besides, many dried plants, gathered in South America by Houston,' says Linnæus, and he adds, 'The English are certainly the most generous people on earth.'

There is still a flavour of poetry about Chelsea, as if poets and philosophers had always dwelt there and left their impress on the place. There are memories too of Tudor sovereigns, for it was during several reigns the resort of the Court and fashion, and 'Ye Old Bun House' and Don Saltero and his tavern-museum in Cheyne Walk. There is, in the British Museum, a long printed catalogue of the rarities to be seen at Don Saltero's coffee-house, all in glass cases numbered—a parent of the South Kensington Museum. The Don was a naturalist, so Linnæus would certainly have visited his collection, and not superciliously, like the 'Spectator,' who talks of the ingenious Don Saltero in a pitying tone of banter. Our own generation has known Turner, George Eliot, Rossetti, and Carlyle, and many other lights, on Chelsea river bank.

The very public-houses at Chelsea and Fulham still have the pretty signs they had in Linnæus's time—the 'Hand and Flower,' 'the Rising Sun,' the 'Daffodil,' the 'Brown Cow,' the 'Three Jolly Gardeners,' the 'World's End,' &c.; and various Puritan sentiments, as 'God encompasseth us' and others; besides Morland's own painting of 'Ye Goat in Boots.' Much of this poetic odour has vanished with the departed Brompton stocks.

Carlyle describes Chelsea as a singular heterogeneous kind of spot, very dirty and confused in some places, quite beautiful in others, abounding with antiquities and the traces of great men—Sir Thomas More, Steele, Smollett, &c. 'Picture a parade running along the shore of the river; a broad highway with huge shady trees; boats lying moored, and a smell of shipping and tan; Battersea Bridge (of wood) a few yards off; the broad river, with white-trousered, white-shirted Cockneys dashing by like arrows in their long canoes of boats; beyond, the green beautiful hills of Surrey with their villages; on the whole a most artificial, green-painted, yet lively, fresh, almost opera-looking business, such as you can fancy. Stroll on the bank of the river and see white-shirted Cockneys in their green canoes, or old pensioners pensively smoking tobacco.' Having become fashionable again with its new embankment, bridges, and its rose-red houses, Chelsea is now more heterogeneous than ever; a greater mixture of blackness and brightness, of squalor and wealth.

Linnæus went to Wimbledon Common and to Kew —pretty, peaceful, still graceful, retired and courtly Kew, the village of dowagers, with its lanes of elms, and soft river scenes with mildly-active life upon them, of a well-to-do pleasuring sort, where the tender grace of a day that has fled seems ever to come back to us. It is like the summer evening of life—of modern life; its apple-blossomed market gardens gone, but all its learning manifest in the magnificent scientific garden lying close behind its river. Kew Church and the youthful games on the green fill up the picture.[1]

The first time Linnæus crossed Putney Heath the sight of the gorse blossom in its blaze of May made him fall on his knees in rapture to thank God for making anything so beautiful.[2]

Impulsiveness like this must have astounded the more stolid Englishman who saw the action. Foreigners, and particularly the Swedes, are more impulsive in their movements than we are. An action of this sort with us would be set down to extravagance or affectation. But the vivacious Linnæus, excited by the long unfamiliar fresh breeze of the heathland—he, a stranger, and well-nigh a dumb creature, felt that the flowers were his friends: they spoke the language that he knew. He was touched by the sight of this flowery wilderness as many persons are moved by hearing grand organ music in a foreign cathedral, where one can without remark indulge the feeling of rapturous thankfulness to the Creator for making so exquisite an universal language. If music spoke not to Linnæus it was not that the grand music of his time was weak in the impressions it could make: it was that Linnæus 'had not

---

[1] There is a caricature portrait of Linnæus in the Museum at Kew, said to have been drawn from life.

[2] It should be remembered that our common furze is entirely confined to Western Europe, and Linnæus had possibly never seen it before.

the sensibility to perceive them.' Perhaps it is the English lack of ready sympathy that causes what struck Dr. Arnold so strongly in travelling abroad—'the total isolation of England from the European world.' Hence one writer has found this only to record of Linnæus in England: 'Of his observations nothing is preserved but the tradition of his rapture at the golden bloom of the furze on Putney Heath.' Did he think the gorse at Putney like glorified juniper buds of his own Småland commons? They should have told him of the old saying that 'when gorse is out of season kissing is out of fashion,' and that is—never. Though the golden flood of the Maybloom is brightest and sweetest, there are nearly always some blossoms to be seen upon the gorse. Linnæus was always an admirer of the furze, and vainly tried to preserve it through a Swedish winter in his greenhouse. The most likely explanation of his extreme emotion lies in the fact of a great resemblance of Putney Heath to some parts of his native country; not Småland itself, but some of the Upland scenery near Upsala. The numerous wind-tost birches and the heather-clad waves of common-land cause a great similarity of landscape.

After his being so long used to the flat fens and sluggish airs of Holland the fresh perfumed breeze playing direct from the Surrey hills would in any case have caused an intoxication of rapture and more excitement than the sight of the gorse itself.

Linnæus and the Gorse.[1]

Over the heath the golden gorse is glowing,
    And making glad the breeze;
And lo! a traveller by the wayside going
    Falls low upon his knees,
And thanks his God for such a glorious vision,
    And such a rich perfume,
As met him in what seemed a dream Elysian,
    Far from his northern home.

So felt the great Linnæus, when before him
    The yellow gorse spread out.
We may not from his far-off grave restore him
    With us to roam about;
But we may drink in, too, that loving spirit
    Which made him seek and find,
Even in the humblest flower that grows, a merit
    Hid from the common mind.

And we, like him, in loving faith may linger
    On many a foreign shore,
Tracing the touch of an Almighty finger
    In plants unknown before;
And, while the beauty of creation feeling,
    Filled with a new delight,
Our hearts, before the great Creator kneeling,
    May bless Him for the sight.—EMILY CARRINGTON.

Sweden does not, for all its distance off, feel to us so foreign a place as Holland, Belgium, or France. The Scandinavians are more like ourselves. I dare say Linnæus felt the same in England as we do in Scandinavia. Gladstone says, 'I do not know whether in any foreign land I ever felt so much at home as in Norway.' He was touched by the universal kindness of the people. We all feel thus in Scandinavia. In Norway we are

[1] *Aunt Judy's Magazine.*

among good-hearted people of our own family, as it were; in Sweden we are in polite yet friendly company. I hardly know which I found the more pleasant: there is a charm in both.

Linnæus had now to present his Oxford credentials, and see what could be gathered for Clifford there, and to try if he might there plant his system. He had, besides, to advertise it personally. These things were not then managed by circulars and letters to the *Times*. Linnæus multiplied himself in travel.

How he must have admired towered Windsor's stately glory and the rich vale of Thames—'England's golden eye'—after the tedious flats of Holland, the wastes of Sweden! He had never seen anything to equal the scenery from Richmond to the spires of Oxford. It was the very opposite to the grandest scenery he knew, and in its way as fine. This was emerald magnificence, that was crystal splendour.

'Hark, the merry Christchurch bells!' These 'agreeable strains of aërial music' were as yet a novelty to him; at Amersfoort only had he heard sweet carillons. There are few carillons in Holland, none in Sweden. Oxford with its domes, spires, and minarets lay before him, 'its rows of shady trees and still monastic edifices in their antique richness and intricate seclusion. 'I never saw a place,' says Haydon the painter, 'that has so much the air of opulence and ease as Oxford. After the bustle, anxieties, fatigue, and harass of a London life, the peace and quiet of those secluded

Gothic-windowed holy chambers of study come over one's feelings with a cooling sensation, as if one had mounted from hell to heaven and been admitted on reprieve from the tortures and fierce passions of the enraged, the malignant, the ignorant, and the lying to the beautiful simplicity of angelic feelings, where all was good, and holy, and pious, and majestic. I need not say it was vacation' (July 16). Learning was livelier on Linnæus's visit in May, and Oxford is always bustling in comparison with Upsala. 'The soil of Oxford is dry, being on a fine gravel. The north is open to cornfields and enclosures for many miles together without a hill to intercept the free current of air.'[1] This wide undulating amphitheatre, filled with spires and towers, must have looked splendid to the young Swede entering by the High Street, the Oxford road from London, or standing upon Maudlin Bridge over the Cherwell.

At Oxford Linnæus was received in a friendly way by Dr. Shaw, who had travelled in Barbary, and who, having read the new system with great pleasure, declared himself his disciple. This was encouraging, but the other Oxford professors were less affable. They were devoted, and with good reason, to the system of Ray—the indefatigable and learned Ray, of whom Dr. Johnson says that he reckoned twenty thousand species of British insects.[2]

---

[1] Old guide-book, dated 1761.
[2] Ray died in 1705, two years before Linnæus's birth; a man of whom England may well be proud.

F. J. Dillenius, a German, born at Darmstadt 1684, the botanical professor at Oxford, received Linnæus haughtily, and with jealousy and dislike, as one who would upset this cherished system of Ray. Dr. Sherard, who was to Dillenius what Clifford was to the Swede, was present at their interview. It is said, 'The English have much to learn from other nations not only in the arts of being serviceable and amiable with grace, but of being so at all.'[1] This is at least equally true of German scientific men.

Dillenius, finding that Linnæus did not understand English, spoke of him before his face to Sherard as 'the young man who would confound all botany.' But Linnæus, though he spoke no foreign tongue, had invented[2] a language, the language of science, reviving a dead language to aid his purpose. Confound and botany being words of Latin origin, Linnæus understood the purport of his observation, though he remained silent for a while. His Swedish politeness was a check on his fiery temper.

Sherard had formerly been consul at Smyrna; he cultivated botany with ardour, discernment, and princely munificence. His vast herbarium and library are among the literary treasures of Oxford.

'The labours of the Sherards and Sir Hans Sloane

---

[1] J. S. Mill.

[2] This must be understood as a figure of speech, as only in the sense of his having given greater precision to the Latin and terms of science can Linnæus be said actually to have *invented* the language of science.

seemed to promise the establishment of the botanic sceptre in England (Chelsea, Eltham); but they were at a standstill for a system.'[1] Ray's as he left it was imperfect, and too complex for general handling. It was because the question was so pressing, 'How is the king's government to be carried on?'—the government, that is, of Sir Hans Sloane—that Linnæus, who had fixed the attention of all Europe with five works, the product, apparently, of a year, causing a revolution of thought through the whole realm of science, was repulsed by them as an innovator and a radical. Dillenius in his letter to Haller treated Linnæus with a moroseness of criticism and harshness of language that the young Swede's learning and endowments did not deserve. Disheartened by his cold reception, and failing to conciliate the professor's kindness, Linnæus called next day to take leave. 'Before I go,' said he, 'I have to request one favour: tell me why you accuse me of confounding botany?'

Dillenius perceived that the youth had understood his remark to Dr. Sherard, but did not care to explain. Linnæus persisted, and the professor produced from the library a part of Linnæus's own 'Genera Plantarum,' printed at Leyden, a copy of which Gronovius had sent to Oxford. Linnæus found N.B. written on almost every page, and was informed that those letters marked the false genera. Linnæus denied this, and they adjourned to the garden. The professor referred to a

[1] Sir J. Smith.

plant which he and other botanists considered to have three stamens. On examination it proved to have only one, as Linnæus had said. 'Oh, it may be so accidentally in a single flower,' said the professor; but on examining a number of them, it was found to be the rule, as Linnæus had stated it. Dillenius, though slow to be convinced, was not above learning truths he did not yet know. He detained Linnæus several days, and promised him what he had before denied—that he should have the plants Clifford was so anxious to procure.[1]

The professor, now somewhat softened, invited the Swede's inspection of his own and the Sherardian collections, and here showed him what would interest him much.

The Linnæan Society possesses a few of Rudbeck's blocks, engraved on rough wood which looks like pine.

At Upsala, as we know, under Rudbeck senior, was laid the foundation of what is justly called 'the great Swedish school of natural history,' when in 1702 a fire reduced almost all the city to ashes, and the works of Rudbeck, with a thousand blocks already cut, and the materials for his work on the natural history of Lapland, were destroyed. All that remained of the great work the 'Campi Elysii,' folio, were a few copies of the second volume, and three only of the first, one of which is in the Sherardian Library at Oxford. The work was planned to be done in twelve volumes. The

[1] This story is given, among the anecdotes related by Linnæus himself, written in Latin, by Dr. Gieseke.

remains were published under the title of 'Reliquiæ Rudbeckianæ,' folio, 1789.

Dillenius also took him more carefully through the botanic garden, which is thus described in the Oxford guide-book of a little later period.[1] 'In the garden are two elegant and useful greenhouses, built by the university for exotics, of which there is as considerable a collection as can be met with anywhere. One of the large aloes, after growing to the height of twenty-one feet, was blown down in 1750. In the quarters within the yew hedges is the greatest variety imaginable of such plants as require no artificial heat to nourish them, all ranged in their proper classes, and numbered. Also two magnificent yew trees, cut in the form of pedestals, but of enormous size, with a flower-pot on the top, and a plant, as it were, growing out of it. The pineapples raised in the hot-house have nearly (!) the same delicious flavour as those raised in warmer climates. The Earl of Danby purchased the ground (containing five acres) of Maudlin College, and gave it to Oxford as a physic garden.' [The gateway is by Inigo Jones.] 'This useful foundation has been much improved by the late Dr. Sherard, who brought from Smyrna a valuable collection of plants. He built and furnished a library for botanical books. One end of this building hath within a few years been altered into a convenient apartment for the professor, whose salary is paid out of the interest of 3,000*l*., given by Dr. Sherard for that

[1] 1761.

purpose. The assistant to the professor is paid by the university.'

Linnæus went to Blenheim with Dillenius, who was now eager for his conversation; also to Ditchley and Stow. They agreed to differ on some points—the merits of Ray, for instance, which Dillenius rightly held to be surpassing; while Linnæus speaks thus jealously of the great English botanist: 'I am at a loss to divine why nobody takes notice of the discoveries of Cæsalpinus, and wishes to ascribe everything to Ray.' This quotation has been badly put into English. Linnæus wrote it in Latin. Yet Linnæus, as he frequently told his pupils in later years, never ceased to esteem Ray, as one of the most penetrating observers of the natural affinity of plants. And this is, after all, the foundation of the natural system—the only lasting one, being based upon nature.

Linnæus, though he cared little for the pictures at Blenheim, having no feeling for works of art, enjoyed the magnificent gardens, and showed himself abundantly interested in nature. He was even then making notes and studies for his 'Flora Anglica,' written in 1754 (eighteen years later), in which he concisely describes the climate of England and its different soils and elevations as favouring the growth of particular plants. He says that Sweden abounds more in alpine, upland, and forest plants than England, which excels in marine plants and such as affect a chalky soil. This English Flora contains nearly a thousand plants, but the mosses and fungi are not introduced.[1] Clifford and Linnæus were not conversant with mosses. The Dutch connoisseurs were all devoted to exotic plants, especially those from America. Linnæus says of himself, 'I do not profess to be even a tyro in mosses. Holland produces very few of this tribe, in which Sweden abounds.' 'Such plants as are not to be found in Sweden are distinguished by the italic type, and of these there are nearly three hundred. A list is subjoined of one hundred, which the author could not fully investigate.'[2]

At Oxford, it has been well said, everything depends upon the society you fall into. If this be uncongenial the place can have no other attractions, besides its scenery, than those of a town full of good libraries. Dr. Arnold quotes the views of Oxford from 'the pretty field,' or from St. John's Gardens, as among the perfectly beautiful scenes in the earth. He was an enthusiast on Thames scenery, particularly specifying that near Oxford—'the streams, the copses, the solitary rock by Bagley Wood, the heights of Shotover, the broken field behind Ferry Hincksey, with its several glimpses of the distant towers and spires.'

Like Dr. Arnold, Linnæus found 'some of the scenes at the junction of the heath country with the rich valley of the Thames very striking,' though doubtless more so from the rich variety of their flora than on account of their merely picturesque aspect. Both of these great

---

[1] Smith.   [2] Ibid.

men delighted to dig up orchis roots in Bagley Wood, and botanise by the wild stream that flows down between Billington and Cowley Marsh.

Linnæus in travelling abroad saw the best, as ordinary tourists abroad see the worst, specimens of humanity—'innkeepers, beggars, touts, and zany Cockneys.' As Dr. Arnold says, in travelling one 'gets an unfavourable impression of the inhabitants in spite of one's judgment.'

What with his explorations, and visits to Professors Martyn and Rand, and Dr. Shaw, and Mrs. Blackburne, another lady botanist, besides an intimate and lasting friendship he contracted with Dr. Isaac Lawson and Mr. Peter Collinson of Mill Hill, near Hendon, one of those cultivated Quakers who are such fervent gardeners—a man of various studies, a friend after his own heart, Linnæus's time passed very agreeably in England, and these friends added many treasures to the store he had procured at Chelsea of the rarest and nondescript plants he was collecting for Clifford.

Linnæus, at Dillenius's request, remained some time at Oxford exploring the country, riding, or oftener walking, round by what is now the Firs, Haddington, and by Livermore, where George Eliot describes 'J. H. Newman's little conventual dwelling, and from whence one gains in returning a fine view of the Oxford towers.' He crossed the original ford whence Oxford took its name, and he saw New College, with its gardens, surrounded by the old city wall, the chapel where William of Wykeham's crosier is kept, and the cloisters, 'which are fine but gloomy, and less beautiful than those of Magdalene,' and the lovely gardens of Merton College.

Perhaps best of all Linnæus loved the Botanical Library, and that glorious Bodleian, whose catalogue he would ransack, and eagerly scan the backs of the books; for the good reason Dr. Johnson gives us—' Knowledge is of two kinds. We know a subject ourselves, or we know where we can find information upon it.'

Linnæus here found out, to his surprise, that Le Vaillant was not actually the first to clearly see the sexes of plants, although he had been given the credit of it. That discovery belongs to Sir Thomas Millington, of Oxford, in the seventeenth century. He flourished about 1670.

Dillenius, writing to Dr. Richardson, says Linnæus stayed eight days at Oxford. This was probably on a first visit, for we hear he remained at Oxford a month altogether.

Pacing the cloistered Gothic arch of Trinity lime grove, the scent of the linden blossom recalled powerfully to his mind the lime tree of his native place. Walking in the physic garden reminded him still more keenly of his Northern home.

Linnæus remained at Oxford till Midsummer Day, when the ceremonial much reminded him of Sweden, and revived with extra force his longing to get home.

The old guide-book before quoted says: 'It is customary on St. John Baptist's Day to have the university sermon preached in the stone pulpit at the south-east corner of the first court within the college gate, the court on that occasion being decked with boughs and strewed with rushes, alluding to St. John's preaching in the wilderness and in commemoration of the hospitals being dedicated to St. John. The most advantageous view of the tower and niche with the head of John Baptist is from the Physick Garden. The tower contains a very musical peal of ten bells. On May Day morning the clerks and choristers assemble on the top of it, and instead of a mass of requiem for King Henry VII. sing cheerful songs and catches.'

Music from the tops of towers is sometimes heard in Scandinavia to this day. I have heard Luther's hymn, and other solemn music, played on trumpets by men standing up outside the central tower of Röskilde Cathedral, on Whit Sunday after morning service. It was sweet to sit in the cathedral square and listen to the aërial music being played there aloft.

> Within the circle of your incantation
> No blight nor mildew falls;
> Nor fierce unrest, nor lust, nor low ambition
> Passes those airy walls.[1]

Dillenius, though he never publicly adopted the Linnæan arrangement, at length became so partial to the young Swede that he would not leave him for an

---

[1] *The Angelus*, Bret Harte.

hour. He was so impressed with his talents that he urged him to reside at Oxford and share the profits of his professorship with him. Dillenius frequently visited Eltham in Kent, where Dr. J. Sherard (William Sherard's brother) had a house and garden. We are not told that Linnæus ever accompanied him thither, but several hints make me opine that he did so. William Sherard's aim was the continuation of Bauhin's 'Pinax.' 'Such assistance as his,' Dillenius said to J. Sherard, speaking of Linnæus, ' in the continuation of Sherard's " Pinax" would be invaluable.' 'The nature of this Pinax,' Pulteney too rashly takes for granted, 'is too well-known to be explained.' Most people do not know what a pinax is; few dictionaries or cyclopædias even give the word—' yea, I know it but in two.' Pinax, or synopsis—*pinox*, a picture of the vegetable kingdom.

' It was undertaken by Sherard as a continuation of Bauhin's " Pinax Theatri Botanici," and it afterwards devolved on Dillenius to carry it forward in a similar manner. No part of it, however, ever came to the press; but the whole MS., preserved in the Botanical Library at Oxford, deserves to be considered as an interesting monument of the scientific industry and erudition of Sherard and his first professor.'

'When I was at Oxford,' writes Linnæus to Haller from Hartecamp, April 1737, ' Dillenius was finishing the " Phytopinax" of Sherard, of which he had then entirely completed the fourth part.' 'Had Sir Hans Sloane been warmer Linnæus might have obtained an establishment in England, which it has been thought was his wish.'[1] As it was, he did not care to stay there helping Dillenius out of his impossibilities.

Homesickness had come on—always a real malady with the Swedes and Swiss—and he could no longer combat the thought of his Elizabeth waiting for him in far-off Falun; so notwithstanding a warmth of friendship on the part of the professors which no one would have expected from the coolness of his welcome, Linnæus left Oxford, and soon afterwards returned to Holland.

Linnæus dedicated his 'Critica Botanica' to Dillenius, of whom, in writing to Haller, he said, 'Dillenius was the only person then in England who cared about or understood the genera of plants.'

Linnæus returned down the quiet silvery Thames to London. Even then the west end of the river presented a remarkable contrast to the 'immensity of London at the pool;' but the full impression of mighty London's vastness never fully struck Linnæus till he was leaving it by the Thames.

'London is the heart of the world, wonderful only from the mass of human beings. No one has any *knowledge* of London in which he lives. It is a huge aggregate of little systems, each of which is again a

[1] Turton.

small anarchy, the members of which do not work together, but scramble against each other.'[1] Linnæus felt this: the Swedes do not shove. Yet this saying of Carlyle's is not altogether true. Each individual wheel of the mighty clock keeps its own round. The philosopher himself admits it. 'The baker's boy brings muffins to the window at a fixed hour every day, and that is all the Londoner knows or wishes to know on the subject. But it turned out good men.'[2] London at once elevates and humbles us. Man is not merely an unit: he is subdivided. Carlyle goes on to say, 'All London-born men, without exception, seem to me narrow-built, considerably perverted men, rather fractions of a man. Hunt, by nature a *very* clever man, is one instance; Mill, in quite another manner, is another.' But it is in their work they are most subdivided. To make the fraction of a pin perfectly is the aim of modern life.

In London lies 'that medley of experience of everything, great and little, which a man can scarcely have anywhere but in the capital': the very opposite of Swedish life, which holds much that we leave out, and knows little of what we most prize—art, wealth, &c. Without art our life were one sordid delirium.

In 1793 Sir J. Smith, first president of the Linnæan Society, set out for Holland to expatiate on ruined Hartecamp, &c., 'after many an anxious look at the lofty plane and cedar trees of Chelsea gardens still

[1] Carlyle. [2] Ibid

waving in unpropitious direction.' Linnæus now looked at them anxiously likewise.

The Chelsea planes and cedars at length showing a favourable wind, after a final week in London, Linnæus sailed with the tide, winding slowly through the forest of masts, the great English forest—a marvellous contrast to the silent pine-forests of Sweden—in the Thames. 'The giant bustle, the coal-heavers, the bargemen, the ten thousand times ten thousand sounds and movements of that monstrous harbour, formed the grandest object I had ever witnessed. One man seems a drop in the ocean: you feel annihilated in the immensity of that heart of all the earth.'[1] And yet a few great minds dominate it all.

In England Linnæus had his mind opened wider than he expected, and in other lines. He knew of Boulton and Watt—at least as Wedgwood, the 'father of the Potteries,' who was so closely allied with them, took a great interest in the modelling of his portrait, we may safely conclude he knew of those iron chieftains who sold *power*.[2] 'What a giant was Watt! fit to stand beside Gutenberg and Columbus as one of the few whose single discoveries have changed the whole course of human civilisation.'[3] Linnæus was one of the very few men in those days capable of duly estimating any form of genius outside poetry, government, or military art. Other literary men were too closely

---

[1] Carlyle.   [2] Dr. Johnson.
[3] Frederic Harrison.

wrapped up in elegant literature to be able to comprehend that men of science and men of business could revolutionise a world.

Linnæus returned to Holland deeply impressed with the importance of England as a country well fitted to forward the interests of natural science.[1]

[1] Sir W. Jardine.

END OF THE FIRST VOLUME.

PRINTED BY
SPOTTISWOODE AND CO., NEW-STREET SQUARE
LONDON

OCTOBER 1886.

GENERAL LISTS OF WORKS

PUBLISHED BY
Messrs. LONGMANS, GREEN, & CO.
39 PATERNOSTER ROW, LONDON, E.C.

HISTORY, POLITICS, HISTORICAL MEMOIRS, &c.

Arnold's Lectures on Modern History. 8vo. 7s. 6d.
Bagwell's Ireland under the Tudors. Vols. 1 and 2. 2 vols. 8vo. 32s.
Beaconsfield's (Lord) Speeches, edited by Kebbel. 2 vols. 8vo. 32s.
Boultbee's History of the Church of England, Pre-Reformation Period. 8vo. 15s.
Buckle's History of Civilisation. 3 vols. crown 8vo. 24s.
Burrows's History of the Family of Brocas of Beaurepaire and Roche Court. Royal 8vo. 42s.
Cox's (Sir G. W.) General History of Greece. Crown 8vo. Maps, 7s. 6d.
Creighton's History of the Papacy during the Reformation. 8vo. Vols. 1 and 2, 32s. Vols. 3 and 4, 24s.
De Tocqueville's Democracy in America. 2 vols. crown 8vo. 16s.
Doyle's English in America: Virginia, Maryland, and the Carolinas, 8vo. 18s.
— — — The Puritan Colonies, 2 vols. 8vo. 36s.
Epochs of Ancient History :—
    Beesly's Gracchi, Marius, and Sulla, 2s. 6d.
    Capes's Age of the Antonines, 2s. 6d.
    — Early Roman Empire, 2s. 6d.
    Cox's Athenian Empire, 2s. 6d.
    — Greeks and Persians, 2s. 6d.
    Curteis's Rise of the Macedonian Empire, 2s. 6d.
    Ihne's Rome to its Capture by the Gauls, 2s. 6d.
    Merivale's Roman Triumvirates, 2s. 6d.
    Sankey's Spartan and Theban Supremacies, 2s. 6d.
    Smith's Rome and Carthage, the Punic Wars, 2s. 6d.
Epochs of Modern History :—
    Church's Beginning of the Middle Ages, 2s. 6d.
    Cox's Crusades, 2s. 6d.
    Creighton's Age of Elizabeth, 2s. 6d.
    Gairdner's Houses of Lancaster and York, 2s. 6d.
    Gardiner's Puritan Revolution, 2s. 6d.
    — Thirty Years' War, 2s. 6d.
    — (Mrs.) French Revolution, 1789-1795, 2s. 6d.
    Hale's Fall of the Stuarts, 2s. 6d.
    Johnson's Normans in Europe, 2s. 6d.
    Longman's Frederick the Great and the Seven Years' War, 2s. 6d.
    Ludlow's War of American Independence, 2s. 6d.
    M'Carthy's Epoch of Reform, 1830-1850, 2s. 6d.
    Morris's Age of Queen Anne, 2s. 6d.
    — The Early Hanoverians, 2s. 6d.
    Seebohm's Protestant Revolution, 2s. 6d.
    Stubbs's Early Plantagenets, 2s. 6d.
    Warburton's Edward III., 2s. 6d.
Epochs of Church History :—
    Overton's The Evangelical Revival in the Eighteenth Century. [2s. 6d.
    Perry's The Reformation in England, 2s. 6d.
    Tucker's The English Church in other Lands, 2s. 6d.
    *** *Other Volumes in preparation.*

London : LONGMANS, GREEN, & CO.

Freeman's Historical Geography of Europe. 2 vols. 8vo. 31s. 6d.
Froude's English in Ireland in the 18th Century. 3 vols. crown 8vo. 18s.
— History of England. Popular Edition. 12 vols. crown 8vo. 3s. 6d. each.
Gardiner's History of England from the Accession of James I. to the Outbreak of the Civil War. 10 vols. crown 8vo. 60s.
— History of the Great Civil War, 1642-1649 (3 vols.) Vol. 1, 1642-1644, 8vo. 21s.
— Outline of English History, B.C. 55-A.D. 1880. Fcp. 8vo. 2s. 6d.
Greville's Journal of the Reign of Queen Victoria, 1837-1852. 3 vols. 8vo. 36s.
Hickson's Ireland in the Seventeenth Century. 2 vols. 8vo. 28s.
Lecky's History of England in the Eighteenth Century. Vols. 1 & 2, 1700-1760, 8vo. 36s. Vols. 3 & 4, 1760-1784, 8vo. 36s.
— History of European Morals. 2 vols. crown 8vo. 16s.
— — Rationalism in Europe. 2 vols. crown 8vo. 16s.
Longman's Lectures on the History of England. 8vo. 15s.
— Life and Times of Edward III. 2 vols. 8vo. 28s.
Macaulay's Complete Works. Library Edition. 8 vols. 8vo. £5. 5s.
— — — Cabinet Edition. 16 vols. crown 8vo. £4. 16s.
— History of England :—
Student's Edition. 2 vols. cr. 8vo. 12s. | Cabinet Edition. 8 vols. post 8vo. 48s.
People's Edition. 4 vols. cr. 8vo. 16s. | Library Edition. 5 vols. 8vo. £4.
Macaulay's Critical and Historical Essays, with Lays of Ancient Rome In One Volume :—
Authorised Edition. Cr. 8vo. 2s. 6d. | Popular Edition. Cr. 8vo. 2s. 6d.
or 3s. 6d. gilt edges. |
Macaulay's Critical and Historical Essays :—
Student's Edition. 1 vol. cr. 8vo. 6s. | Cabinet Edition. 4 vols. post 8vo. 24s.
People's Edition. 2 vols. cr. 8vo. 8s. | Library Edition. 3 vols. 8vo. 36s.
Macaulay's Speeches corrected by Himself. Crown 8vo. 3s. 6d.
Malmesbury's (Earl of) Memoirs of an Ex-Minister. Crown 8vo. 7s. 6d.
Maxwell's (Sir W. S.) Don John of Austria. Library Edition, with numerous Illustrations. 2 vols. royal 8vo. 42s.
May's Constitutional History of England, 1760-1870. 3 vols. crown 8vo. 18s.
— Democracy in Europe. 2 vols. 8vo. 32s.
Merivale's Fall of the Roman Republic. 12mo. 7s. 6d.
— General History of Rome, B.C. 753-A.D. 476. Crown 8vo. 7s. 6d.
— History of the Romans under the Empire. 8 vols. post 8vo. 48s.
Nelson's (Lord) Letters and Despatches. Edited by J. K. Laughton. 8vo. 16s.
Outlines of Jewish History from B.C. 586 to C.E. 1885. By the author of 'About the Jews since Bible Times.' Fcp. 8vo. 3s. 6d.
Pears' The Fall of Constantinople. 8vo. 16s.
Seebohm's Oxford Reformers—Colet, Erasmus, & More. 8vo. 14s.
Short's History of the Church of England. Crown 8vo. 7s. 6d.
Smith's Carthage and the Carthaginians. Crown 8vo. 10s. 6d.
Taylor's Manual of the History of India. Crown 8vo. 7s. 6d.
Walpole's History of England, 1815-1841. Vols. 1-3, 8vo. £2. 14s. Vols. 4 & 5, 8vo. 36s.
Wylie's History of England under Henry IV. Vol. 1, crown 8vo. 10s. 6d.

London : LONGMANS, GREEN, & CO.

BIOGRAPHICAL WORKS.

Armstrong's (E. J.) Life and Letters. Edited by G. F. Armstrong. Fcp. 8vo. 7s. 6d.
Bacon's Life and Letters, by Spedding. 7 vols. 8vo. £4. 4s.
Bagehot's Biographical Studies. 1 vol. 8vo. 12s.
Carlyle's Life, by J. A. Froude. Vols. 1 & 2. 1795–1835, 8vo. 32s. Vols. 3 & 4, 1834–1881, 8vo. 32s.
— (Mrs.) Letters and Memorials. 3 vols. 8vo. 36s.
De Witt (John) Life of, by A. C. Pontalis. Translated. 2 vols. 8vo. 36s.
Doyle (Sir F. H.) Reminiscences and Opinions. 8vo. 16s.
English Worthies. Edited by Andrew Lang. Crown 8vo. 2s. 6d. each.
  Charles Darwin. By Grant Allen. | Marlborough. By George Saintsbury.
  Shaftesbury (The First Earl). By | Steele. By Austin Dobson.
   H. D. Traill. | Ben Jonson. By J. A. Symonds.
  Admiral Blake. By David Hannay. |
  ** *Other Volumes in preparation.*
Fox (Charles James) The Early History of. By Sir G. O. Trevelyan, Bart. Crown 8vo. 6s.
Froude's Cæsar: a Sketch. Crown 8vo. 6s.
Hamilton's (Sir W. R.) Life, by Graves. Vols. 1 and 2, 8vo. 15s. each.
Havelock's Life, by Marshman. Crown 8vo. 3s. 6d.
Hobart Pacha's Sketch of my Life. Crown 8vo. 7s. 6d.
Hullah's (John) Life. By his Wife. Crown 8vo. 6s.
Macaulay's (Lord) Life and Letters. By his Nephew, Sir G. O. Trevelyan, Bart., M.P. Popular Edition, 1 vol. crown 8vo. 6s. Cabinet Edition, 2 vols. post 8vo. 12s. Library Edition, 2 vols. 8vo. 36s.
Mendelssohn's Letters. Translated by Lady Wallace. 2 vols. cr. 8vo. 5s. each.
Mill (James) Biography of, by Prof. Bain. Crown 8vo. 5s.
— (John Stuart) Recollections of, by Prof. Bain. Crown 8vo. 2s. 6d.
— — Autobiography. 8vo. 7s. 6d.
Mozley's Reminiscences of Oriel College. 2 vols. crown 8vo. 18s.
— — Towns, Villages, and Schools. 2 vols. cr. 8vo. 18s.
Müller's (Max) Biographical Essays. Crown 8vo. 7s. 6d.
Newman's Apologia pro Vitâ Suâ. Crown 8vo. 6s.
Pasteur (Louis) His Life and Labours. Crown 8vo. 7s. 6d.
Shakespeare's Life (Outlines of), by Halliwell-Phillipps. 2 vols. royal 8vo. 10s. 6d.
Southey's Correspondence with Caroline Bowles. 8vo. 14s.
Stephen's Essays in Ecclesiastical Biography. Crown 8vo. 7s. 6d.
Taylor's (Sir Henry) Autobiography. 2 vols. 8vo. 32s.
Wellington's Life, by Gleig. Crown 8vo. 6s.

MENTAL AND POLITICAL PHILOSOPHY, FINANCE, &c.

Amos's View of the Science of Jurisprudence. 8vo. 18s.
— Primer of the English Constitution. Crown 8vo. 6s.
Bacon's Essays, with Annotations by Whately. 8vo. 10s. 6d.
— Works, edited by Spedding. 7 vols. 8vo. 73s. 6d.
Bagehot's Economic Studies, edited by Hutton. 8vo. 10s. 6d.
— The Postulates of English Political Economy. Crown 8vo. 2s. 6d.
Bain's Logic, Deductive and Inductive. Crown 8vo. 10s. 6d.
  PART I. Deduction, 4s. | PART II. Induction, 6s. 6d.
— Mental and Moral Science. Crown 8vo. 10s. 6d.
— The Senses and the Intellect. 8vo. 15s.
— The Emotions and the Will. 8vo. 15s.
— Practical Essays. Crown 8vo. 4s. 6d.

London : LONGMANS, GREEN, & CO.

Buckle's (H. T.) Miscellaneous and Posthumous Works. 2 vols. crown 8vo. 21s.
Crozier's Civilization and Progress. 8vo. 14s.
Crump's A Short Enquiry into the Formation of English Political Opinion. 8vo. 7s. 6d.
Dowell's A History of Taxation and Taxes in England. 4 vols. 8vo. 48s.
Green's (Thomas Hill) Works. (3 vols.) Vols. 1 & 2, Philosophical Works. 8vo. 16s. each.
Hume's Essays, edited by Green & Grose. 2 vols. 8vo. 28s.
— Treatise of Human Nature, edited by Green & Grose. 2 vols. 8vo. 28s.
Lang's Custom and Myth : Studies of Early Usage and Belief. Crown 8vo. 7s. 6d.
Leslie's Essays in Political and Moral Philosophy. 8vo. 10s. 6d.
Lewes's History of Philosophy. 2 vols. 8vo. 32s.
Lubbock's Origin of Civilisation. 8vo. 18s.
Macleod's Principles of Economical Philosophy. In 2 vols. Vol. 1, 8vo. 15s. Vol. 2, Part I. 12s.
— The Elements of Economics. (2 vols.) Vol. 1, cr. 8vo. 7s. 6d. Vol. 2, Part I. cr. 8vo. 7s. 6d.
— The Elements of Banking. Crown 8vo. 5s.
— The Theory and Practice of Banking. Vol. 1, 8vo. 12s. Vol. 2, 14s.
— Elements of Political Economy. 8vo. 16s.
— Economics for Beginners. 8vo. 2s. 6d.
— Lectures on Credit and Banking. 8vo. 5s.
Mill's (James) Analysis of the Phenomena of the Human Mind. 2 vols. 8vo. 28s.
Mill (John Stuart) on Representative Government. Crown 8vo. 2s.
— — on Liberty. Crown 8vo. 1s. 4d.
— — Examination of Hamilton's Philosophy. 8vo. 16s.
— — Logic. Crown 8vo. 5s.
— — Principles of Political Economy. 2 vols. 8vo. 30s. People's Edition, 1 vol. crown 8vo. 5s.
— — Subjection of Women. Crown 8vo. 6s.
— — Utilitarianism. 8vo. 5s.
— — Three Essays on Religion, &c. 8vo. 5s.
Mulhall's History of Prices since 1850. Crown 8vo. 6s.
Sandars's Institutes of Justinian, with English Notes. 8vo. 18s.
Seebohm's English Village Community. 8vo. 16s.
Sully's Outlines of Psychology. 8vo. 12s. 6d.
— Teacher's Handbook of Psychology. Crown 8vo. 6s. 6d.
Swinburne's Picture Logic. Post 8vo. 5s.
Thompson's A System of Psychology. 2 vols. 8vo. 36s.
Thomson's Outline of Necessary Laws of Thought. Crown 8vo. 6s.
Twiss's Law of Nations in Time of War. 8vo. 21s.
— — In Time of Peace. 8vo. 15s.
Webb's The Veil of Isis. 8vo. 10s. 6d.
Whately's Elements of Logic. Crown 8vo. 4s. 6d.
— — — Rhetoric. Crown 8vo. 4s. 6d.
Wylie's Labour, Leisure, and Luxury. Crown 8vo. 6s.
Zeller's History of Eclecticism in Greek Philosophy. Crown 8vo. 10s. 6d.
— Plato and the Older Academy. Crown 8vo. 18s.
— Pre-Socratic Schools. 2 vols. crown 8vo. 30s.
— Socrates and the Socratic Schools. Crown 8vo. 10s. 6d.
— Stoics, Epicureans, and Sceptics. Crown 8vo. 15s.
— Outlines of the History of Greek Philosophy. Crown 8vo. 10s. 6d.

London : LONGMANS, GREEN, & CO.

MISCELLANEOUS WORKS.

A. K. H. B., The Essays and Contributions of. Crown 8vo.
    Autumn Holidays of a Country Parson. 3s. 6d.
    Changed Aspects of Unchanged Truths. 3s. 6d.
    Common-Place Philosopher in Town and Country. 3s. 6d.
    Critical Essays of a Country Parson. 3s. 6d.
    Counsel and Comfort spoken from a City Pulpit. 3s. 6d.
    Graver Thoughts of a Country Parson. Three Series. 3s. 6d. each.
    Landscapes, Churches, and Moralities. 3s. 6d.
    Leisure Hours in Town. 3s. 6d. Lessons of Middle Age. 3s. 6d.
    Our Little Life. Essays Consolatory and Domestic. Two Series. 3s. 6d.
    Present-day Thoughts. 3s. 6d.   [each.
    Recreations of a Country Parson. Three Series. 3s. 6d. each.
    Seaside Musings on Sundays and Week-Days. 3s. 6d.
    Sunday Afternoons in the Parish Church of a University City. 3s. 6d.
Armstrong's (Ed. J.) Essays and Sketches. Fcp. 8vo. 5s.
Arnold's (Dr. Thomas) Miscellaneous Works. 8vo. 7s. 6d.
Bagehot's Literary Studies, edited by Hutton. 2 vols. 8vo. 28s.
Beaconsfield (Lord), The Wit and Wisdom of. Crown 8vo. 1s. boards; 1s. 6d. cl.
Evans's Bronze Implements of Great Britain. 8vo. 25s.
Farrar's Language and Languages. Crown 8vo. 6s.
Froude's Short Studies on Great Subjects. 4 vols. crown 8vo. 24s.
Lang's Letters to Dead Authors. Fcp. 8vo. 6s. 6d.
Macaulay's Miscellaneous Writings. 2 vols. 8vo. 21s. 1 vol. crown 8vo. 4s. 6d.
    — Miscellaneous Writings and Speeches. Crown 8vo. 6s.
    — Miscellaneous Writings, Speeches, Lays of Ancient Rome, &c. Cabinet Edition. 4 vols. crown 8vo. 24s.
    — Writings, Selections from. Crown 8vo. 6s.
Müller's (Max) Lectures on the Science of Language. 2 vols. crown 8vo. 16s.
    — — Lectures on India. 8vo. 12s. 6d.
Smith (Sydney) The Wit and Wisdom of. Crown 8vo. 1s. boards; 1s. 6d. cloth.
Wilkinson's The Friendly Society Movement. Crown 8vo. 2s. 6d.

ASTRONOMY.

Herschel's Outlines of Astronomy. Square crown 8vo. 12s.
Neison's Work on the Moon. Medium 8vo. 31s. 6d.
Proctor's Larger Star Atlas. Folio, 15s. or Maps only, 12s. 6d.
    — New Star Atlas. Crown 8vo. 5s.
    — Light Science for Leisure Hours. 3 Series. Crown 8vo. 5s. each.
    — The Moon. Crown 8vo. 6s.
    — Other Worlds than Ours. Crown 8vo. 5s.
    — The Sun. Crown 8vo. 14s.
    — Studies of Venus-Transits. 8vo. 5s.
    — Orbs Around Us. Crown 8vo. 5s.
    — Universe of Stars. 8vo. 10s. 6d.
Webb's Celestial Objects for Common Telescopes. Crown 8vo. 9s.

THE 'KNOWLEDGE' LIBRARY.
Edited by RICHARD A. PROCTOR.

How to Play Whist. Crown 8vo. 5s.
Home Whist. 16mo. 1s.
The Borderland of Science. Cr. 8vo. 6s.
Nature Studies. Crown 8vo. 6s.
Leisure Readings. Crown 8vo. 6s.
The Stars in their Seasons. Imp. 8vo. 5s.
Myths and Marvels of Astronomy. Crown 8vo. 6s.
Pleasant Ways in Science. Cr. 8vo. 6s.
Star Primer. Crown 4to. 2s. 6d.
The Seasons Pictured. Demy 4to. 5s.
Strength and Happiness. Cr. 8vo. 5s.
Rough Ways made Smooth. Cr. 8vo. 6s.
The Expanse of Heaven. Cr. 8vo. 5s.
Our Place among Infinities. Cr. 8vo. 5s.

London: LONGMANS, GREEN, & CO.

CLASSICAL LANGUAGES AND LITERATURE.

Æschylus, The Eumenides of. Text, with Metrical English Translation, by J. F. Davies. 8vo. 7s.
Aristophanes' The Acharnians, translated by R. Y. Tyrrell. Crown 8vo. 2s. 6d.
Aristotle's The Ethics, Text and Notes, by Sir Alex. Grant, Bart. 2 vols. 8vo. 32s.
— The Nicomachean Ethics, translated by Williams, crown 8vo. 7s. 6d.
— The Politics, Books I. III. IV. (VII.) with Translation, &c. by Bolland and Lang. Crown 8vo. 7s. 6d.
Becker's *Charicles* and *Gallus*, by Metcalfe. Post 8vo. 7s. 6d. each.
Cicero's Correspondence, Text and Notes, by R. Y. Tyrrell. Vols. 1 & 2, 8vo. 12s. each.
Homer's Iliad, Homometrically translated by Cayley. 8vo. 12s. 6d.
— — Greek Text, with Verse Translation, by W. C. Green. Vol. 1, Books I.-XII. Crown 8vo. 6s.
Mahaffy's Classical Greek Literature. Crown 8vo. Vol. 1, The Poets, 7s. 6d. Vol. 2, The Prose Writers, 7s. 6d.
Plato's Parmenides, with Notes, &c. by J. Maguire. 8vo. 7s. 6d.
Virgil's Works, Latin Text, with Commentary, by Kennedy. Crown 8vo. 10s. 6d.
— Æneid, translated into English Verse, by Conington. Crown 8vo. 9s.
— — — — — by W. J. Thornhill. Cr. 8vo. 7s. 6d.
— Poems, — — — Prose, by Conington. Crown 8vo. 9s.
Witt's Myths of Hellas, translated by F. M. Younghusband. Crown 8vo. 3s. 6d.
— The Trojan War, — — Fcp. 8vo. 2s.
— The Wanderings of Ulysses, — Crown 8vo. 3s. 6d.

NATURAL HISTORY, BOTANY, & GARDENING.

Allen's Flowers and their Pedigrees. Crown 8vo. Woodcuts, 5s.
Decaisne and Le Maout's General System of Botany. Imperial 8vo. 31s. 6d.
Dixon's Rural Bird Life. Crown 8vo. Illustrations, 5s.
Hartwig's Aerial World, 8vo. 10s. 6d.
— Polar World, 8vo. 10s. 6d.
— Sea and its Living Wonders. 8vo. 10s. 6d.
— Subterranean World, 8vo. 10s. 6d.
— Tropical World, 8vo. 10s. 6d.
Lindley's Treasury of Botany. 2 vols. fcp. 8vo. 12s.
Loudon's Encyclopædia of Gardening. 8vo. 21s.
— — Plants. 8vo. 42s.
Rivers's Orchard House. Crown 8vo. 5s.
— Rose Amateur's Guide. Fcp. 8vo. 4s. 6d.
— Miniature Fruit Garden. Fcp. 8vo. 4s.
Stanley's Familiar History of British Birds. Crown 8vo. 6s.
Wood's Bible Animals. With 112 Vignettes. 8vo. 10s. 6d.
— Common British Insects. Crown 8vo. 3s. 6d.
— Homes Without Hands, 8vo. 10s. 6d.
— Insects Abroad, 8vo. 10s. 6d.
— Horse and Man. 8vo. 14s.
— Insects at Home. With 700 Illustrations. 8vo. 10s. 6d.
— Out of Doors. Crown 8vo. 5s.
— Petland Revisited. Crown 8vo. 7s. 6d.
— Strange Dwellings. Crown 8vo. 5s. Popular Edition, 4to. 6d.

London: LONGMANS, GREEN, & CO.

THE FINE ARTS AND ILLUSTRATED EDITIONS.

Dresser's Arts and Art Manufactures of Japan. Square crown 8vo. 31s. 6d.
Eastlake's Household Taste in Furniture, &c. Square crown 8vo. 14s.
Jameson's Sacred and Legendary Art. 6 vols. square 8vo.
    Legends of the Madonna. 1 vol. 21s.
    — — — Monastic Orders. 1 vol. 21s.
    — — — Saints and Martyrs. 2 vols. 31s. 6d.
    — — — Saviour. Completed by Lady Eastlake. 2 vols. 42s.
Macaulay's Lays of Ancient Rome, illustrated by Scharf. Fcp. 4to. 10s. 6d.
The same, with *Ivry* and the *Armada*, illustrated by Weguelin. Crown 8vo. 3s. 6d.
New Testament (The) illustrated with Woodcuts after Paintings by the Early Masters. 4to. 21s.

CHEMISTRY, ENGINEERING, & GENERAL SCIENCE.

Arnott's Elements of Physics or Natural Philosophy. Crown 8vo. 12s. 6d.
Barrett's English Glees and Part-Songs: their Historical Development. Crown 8vo. 7s. 6d.
Bourne's Catechism of the Steam Engine. Crown 8vo. 7s. 6d.
    — Examples of Steam, Air, and Gas Engines. 4to. 70s.
    — Handbook of the Steam Engine. Fcp. 8vo. 9s.
    — Recent Improvements in the Steam Engine. Fcp. 8vo. 6s.
    — Treatise on the Steam Engine. 4to. 42s.
Buckton's Our Dwellings, Healthy and Unhealthy. Crown 8vo. 3s. 6d.
Clerk's The Gas Engine. With Illustrations. Crown 8vo. 7s. 6d.
Crookes's Select Methods in Chemical Analysis. 8vo. 24s.
Culley's Handbook of Practical Telegraphy. 8vo. 16s.
Fairbairn's Useful Information for Engineers. 3 vols. crown 8vo. 31s. 6d.
    — Mills and Millwork. 1 vol. 8vo. 25s.
Ganot's Elementary Treatise on Physics, by Atkinson. Large crown 8vo. 15s.
    — Natural Philosophy, by Atkinson. Crown 8vo. 7s. 6d.
Grove's Correlation of Physical Forces. 8vo. 15s.
Haughton's Six Lectures on Physical Geography. 8vo. 15s.
Helmholtz on the Sensations of Tone. Royal 8vo. 28s.
Helmholtz's Lectures on Scientific Subjects. 2 vols. crown 8vo. 7s. 6d. each.
Hudson and Gosse's The Rotifera or 'Wheel Animalcules.' With 30 Coloured Plates. 6 parts. 4to. 10s. 6d. each. Complete, 2 vols. 4to. £3. 10s.
Hullah's Lectures on the History of Modern Music. 8vo. 8s. 6d.
    — Transition Period of Musical History. 8vo. 10s. 6d.
Jackson's Aid to Engineering Solution. Royal 8vo. 21s.
Jago's Inorganic Chemistry, Theoretical and Practical. Fcp. 8vo. 2s.
Kerl's Metallurgy, adapted by Crookes and Röhrig. 3 vols. 8vo. £4. 19s.
Kolbe's Short Text-Book of Inorganic Chemistry. Crown 8vo. 7s. 6d.
Lloyd's Treatise on Magnetism. 8vo. 10s. 6d.
Macalister's Zoology and Morphology of Vertebrate Animals. 8vo. 10s. 6d.
Macfarren's Lectures on Harmony. 8vo. 12s.
Miller's Elements of Chemistry, Theoretical and Practical. 3 vols. 8vo. Part I. Chemical Physics, 16s. Part II. Inorganic Chemistry, 24s. Part III. Organic Chemistry, price 31s. 6d.

London: LONGMANS, GREEN, & CO.

Mitchell's Manual of Practical Assaying. 8vo. 31s. 6d.
Northcott's Lathes and Turning. 8vo. 18s.
Owen's Comparative Anatomy and Physiology of the Vertebrate Animals. 3 vols. 8vo. 73s. 6d.
Piesse's Art of Perfumery. Square crown 8vo. 21s.
Reynolds's Experimental Chemistry. Fcp. 8vo. Part I. 1s. 6d. Part II. 2s. 6d. Part III. 3s. 6d.
Schellen's Spectrum Analysis. 8vo. 31s. 6d.
Sennett's Treatise on the Marine Steam Engine. 8vo. 21s.
Smith's Air and Rain. 8vo. 24s.
Stoney's The Theory of the Stresses on Girders, &c. Royal 8vo. 36s.
Swinton's Electric Lighting: Its Principles and Practice. Crown 8vo. 5s.
Tilden's Practical Chemistry. Fcp. 8vo. 1s. 6d.
Tyndall's Faraday as a Discoverer. Crown 8vo. 3s. 6d.
— Floating Matter of the Air. Crown 8vo. 7s. 6d.
— Fragments of Science. 2 vols. post 8vo. 16s.
— Heat a Mode of Motion. Crown 8vo. 12s.
— Lectures on Light delivered in America. Crown 8vo. 5s.
— Lessons on Electricity. Crown 8vo. 2s. 6d.
— Notes on Electrical Phenomena. Crown 8vo. 1s. sewed, 1s. 6d. cloth.
— Notes of Lectures on Light. Crown 8vo. 1s. sewed, 1s. 6d. cloth.
— Sound, with Frontispiece and 203 Woodcuts. Crown 8vo. 10s. 6d.
Watts's Dictionary of Chemistry. 9 vols. medium 8vo. £15. 2s. 6d.
Wilson's Manual of Health-Science. Crown 8vo. 2s. 6d.

THEOLOGICAL AND RELIGIOUS WORKS.

Arnold's (Rev. Dr. Thomas) Sermons. 6 vols. crown 8vo. 5s. each.
Boultbee's Commentary on the 39 Articles. Crown 8vo. 6s.
Browne's (Bishop) Exposition of the 39 Articles. 8vo. 16s.
Bullinger's Critical Lexicon and Concordance to the English and Greek New Testament. Royal 8vo. 15s.
Colenso on the Pentateuch and Book of Joshua. Crown 8vo. 6s.
Conder's Handbook of the Bible. Post 8vo. 7s. 6d.
Conybeare & Howson's Life and Letters of St. Paul:—
   Library Edition, with Maps, Plates, and Woodcuts. 2 vols. square crown 8vo. 21s.
   Student's Edition, revised and condensed, with 46 Illustrations and Maps. 1 vol. crown 8vo. 7s. 6d.
Cox's (Homersham) The First Century of Christianity. 8vo. 12s.
Davidson's Introduction to the Study of the New Testament. 2 vols. 8vo. 30s.
Edersheim's Life and Times of Jesus the Messiah. 2 vols. 8vo. 24s.
— Prophecy and History in relation to the Messiah. 8vo. 12s.
Ellicott's (Bishop) Commentary on St. Paul's Epistles. 8vo. Galatians, 8s. 6d. Ephesians, 8s. 6d. Pastoral Epistles, 10s. 6d. Philippians, Colossians and Philemon, 10s. 6d. Thessalonians, 7s. 6d.
— Lectures on the Life of our Lord. 8vo. 12s.
Ewald's Antiquities of Israel, translated by Solly. 8vo. 12s. 6d.
— History of Israel, translated by Carpenter & Smith. Vols. 1–7, 8vo. £5.
Hobart's Medical Language of St. Luke. 8vo. 16s.

London: LONGMANS, GREEN, & CO.

General Lists of Works.

Hopkins's Christ the Consoler. Fcp. 8vo. 2s. 6d.
Jukes's New Man and the Eternal Life. Crown 8vo. 6s.
— Second Death and the Restitution of all Things. Crown 8vo. 3s. 6d.
— Types of Genesis. Crown 8vo. 7s. 6d.
— The Mystery of the Kingdom. Crown 8vo. 3s. 6d.
Lenormant's New Translation of the Book of Genesis. Translated into English. 8vo. 10s. 6d.
Lyra Germanica: Hymns translated by Miss Winkworth. Fcp. 8vo. 5s.
Macdonald's (G.) Unspoken Sermons. Two Series, Crown 8vo. 3s. 6d. each.
— The Miracles of our Lord. Crown 8vo. 3s. 6d.
Manning's Temporal Mission of the Holy Ghost. Crown 8vo. 8s. 6d.
Martineau's Endeavours after the Christian Life. Crown 8vo. 7s. 6d.
— Hymns of Praise and Prayer. Crown 8vo. 4s. 6d. 32mo. 1s. 6d.
— Sermons, Hours of Thought on Sacred Things. 2 vols. 7s. 6d. each.
Monsell's Spiritual Songs for Sundays and Holidays. Fcp. 8vo. 5s. 18mo. 2s.
Müller's (Max) Origin and Growth of Religion. Crown 8vo. 7s. 6d.
— — Science of Religion. Crown 8vo. 7s. 6d.
Newman's Apologia pro Vitâ Suâ. Crown 8vo. 6s.
— The Idea of a University Defined and Illustrated. Crown 8vo. 7s.
— Historical Sketches. 3 vols. crown 8vo. 6s. each.
— Discussions and Arguments on **Various Subjects.** Crown 8vo. 6s.
— An Essay on the Development of **Christian Doctrine.** Crown 8vo. 6s.
— Certain Difficulties Felt by Anglicans in Catholic Teaching Considered. Vol. 1, crown 8vo. 7s. 6d. Vol. 2, crown 8vo. 5s. 6d.
— The Via Media of the Anglican Church, Illustrated in **Lectures, &c.** 2 vols. crown 8vo. 6s. each
— Essays, Critical and Historical. 2 vols. crown 8vo. 12s.
— Essays on Biblical and on Ecclesiastical Miracles. Crown 8vo. 6s.
— An Essay in Aid of a Grammar of **Assent.** 7s. 6d.
Overton's Life in the English Church (1660–1714). 8vo. 14s.
Rogers's Eclipse of Faith. Fcp. 8vo. 5s.
— Defence of the Eclipse of Faith. Fcp. 8vo. 3s. 6d.
Sewell's (Miss) Night Lessons from Scripture. 32mo. 3s. 6d.
— — **Passing** Thoughts on Religion. Fcp. 8vo. 3s. 6d.
— — Preparation **for the** Holy Communion. 32mo. 3s.
Smith's Voyage and Shipwreck of St. Paul. Crown 8vo. 7s. 6d.
Supernatural Religion. Complete Edition. 3 vols. 8vo. 36s.
Taylor's (Jeremy) Entire Works. With Life by Bishop Heber. **Edited by the** Rev. C. P. Eden. 10 vols. 8vo. £5. 5s.

TRAVELS, ADVENTURES, &c.

Alpine Club (The) Map of Switzerland. In Four **Sheets.** 42s.
Baker's Eight Years in Ceylon. Crown 8vo. 5s.
— Rifle and Hound in Ceylon. Crown 8vo. 5s.
Ball's Alpine Guide. 3 vols. post 8vo. with Maps and Illustrations:—I. Western Alps. 6s. 6d. II. Central Alps. 7s. 6d. III. Eastern Alps, 10s. 6d.
Ball on Alpine Travelling, and on the Geology of the Alps, 1s.
Bent's The Cyclades, or Life among the Insular Greeks. Crown 8vo. 12s. 6d.

London: LONGMANS, GREEN, & CO.

Brassey's Sunshine and Storm in the East. Library Edition, 8vo. 21s. Cabinet Edition, crown 8vo. 7s. 6d.
— Voyage in the Yacht 'Sunbeam.' Library Edition, 8vo. 21s. Cabinet Edition, crown 8vo. 7s. 6d. School Edition, fcp. 8vo. 2s. Popular Edition, 4to. 6d.
— In the Trades, the Tropics, and the 'Roaring Forties.' Édition de Luxe, 8vo. £3. 13s. 6d. Library Edition, 8vo. 21s.
Crawford's Across the Pampas and the Andes. Crown 8vo. 7s. 6d.
Dent's Above the Snow Line. Crown 8vo. 7s. 6d.
Froude's Oceana; or, England and her Colonies. Crown 8vo. 2s. boards; 2s. 6d. cloth.
Hassall's San Remo Climatically considered. Crown 8vo. 5s.
Howitt's Visits to Remarkable Places. Crown 8vo. 7s. 6d.
Three in Norway. By Two of Them. Crown 8vo. Illustrations, 6s.

WORKS OF FICTION.

Beaconsfield's (The Earl of) Novels and Tales. Hughenden Edition, with 2 Portraits on Steel and 11 Vignettes on Wood. 11 vols. crown 8vo. £2. 2s. Cheap Edition, 11 vols. crown 8vo. 1s. each, boards; 1s. 6d. each, cloth.

Lothair.	Contarini Fleming.
Sybil.	Alroy, Ixion, &c.
Coningsby.	The Young Duke, &c.
Tancred.	Vivian Grey.
Venetia.	Endymion.
Henrietta Temple.	

Black Poodle (The) and other Tales. By the Author of 'Vice Versâ.' Cr. 8vo. 6s.
Brabourne's (Lord) Friends and Foes from Fairyland. Crown 8vo. 6s.
Harte (Bret) On the Frontier. Three Stories. 16mo. 1s.
— — By Shore and Sedge. Three Stories. 16mo. 1s.
— — In the Carquinez Woods. Crown 8vo. 2s. boards; 2s. 6d. cloth.
Melville's (Whyte) Novels. 8 vols. fcp. 8vo. 1s. each, boards; 1s. 6d. each, cloth.

Digby Grand.	Good for Nothing.
General Bounce.	Holmby House.
Kate Coventry.	The Interpreter.
The Gladiators.	The Queen's Maries.

Novels by the Author of 'The Atelier du Lys':
The Atelier du Lys; or, An Art Student in the Reign of Terror. Crown 8vo. 2s. 6d.
Mademoiselle Mori: a Tale of Modern Rome. Crown 8vo. 2s. 6d.
In the Olden Time: a Tale of the Peasant War in Germany. Crown 8vo. 2s. 6d.
Hester's Venture. Crown 8vo. 6s.
Oliphant's (Mrs.) Madam. Crown 8vo. 2s. 6d.
— — In Trust: the Story of a Lady and her Lover. Crown 8vo. 2s. boards; 2s. 6d. cloth.
Payn's (James) The Luck of the Darrells. Crown 8vo. 3s. 6d.
— — Thicker than Water. Crown 8vo. 2s. boards; 2s. 6d. cloth.
Reader's Fairy Prince Follow-my-Lead. Crown 8vo. 5s.
Sewell's (Miss) Stories and Tales. Crown 8vo. 1s. each, boards; 1s. 6d. cloth; 2s. 6d. cloth extra, gilt edges.

Amy Herbert. Cleve Hall.	A Glimpse of the World.
The Earl's Daughter.	Katharine Ashton.
Experience of Life.	Laneton Parsonage.
Gertrude. Ivors.	Margaret Percival. Ursula.

London: LONGMANS, GREEN, & CO.

General Lists of Works. 11

Stevenson's (R. L.) The Dynamiter. Fcp. 8vo. 1s. sewed; 1s. 6d. cloth.
— — Strange Case of Dr. Jekyll and Mr. Hyde. Fcp. 8vo. 1s. sewed; 1s. 6d. cloth.
Trollope's (Anthony) Novels. Fcp. 8vo. 1s. each, boards; 1s. 6d. cloth.
    The Warden | Barchester Towers.

POETRY AND THE DRAMA.

Armstrong's (Ed. J.) Poetical Works. Fcp. 8vo. 5s.
— (G. F.) Poetical Works:—
  Poems, Lyrical and Dramatic. Fcp. 8vo. 6s.
  Ugone: a Tragedy. Fcp. 8vo. 6s.
  A Garland from Greece. Fcp. 8vo. 9s.
  King Saul. Fcp. 8vo. 5s.
  King David. Fcp. 8vo. 6s.
  King Solomon. Fcp. 8vo. 6s.
  Stories of Wicklow. Fcp. 8vo. 9s.
Bowen's Harrow Songs and other Verses. Fcp. 8vo. 2s. 6d.; or printed on hand-made paper, 5s.
Bowdler's Family Shakespeare. Medium 8vo. 14s. 6 vols. fcp. 8vo. 21s.
Dante's Divine Comedy, translated by James Innes Minchin. Crown 8vo. 15s.
Goethe's Faust, translated by Birds. Large crown 8vo. 12s. 6d.
— — translated by Webb. 8vo. 12s. 6d.
— — edited by Selss. Crown 8vo. 5s.
Ingelow's Poems. Vols. 1 and 2, fcp. 8vo. 12s. Vol. 3 fcp. 8vo. 5s.
— Lyrical and other Poems. Fcp. 8vo. 2s. 6d. cloth, plain; 3s. cloth, gilt edges.
Macaulay's Lays of Ancient Rome, with Ivry and the Armada. Illustrated by Weguelin. Crown 8vo. 3s. 6d. gilt edges.
The same, Popular Edition. Illustrated by Scharf. Fcp. 4to. 6d. swd., 1s. cloth.
Reader's Voices from Flowerland, a Birthday Book, 2s. 6d. cloth, 3s. 6d. roan.
Southey's Poetical Works. Medium 8vo. 14s.
Stevenson's A Child's Garden of Verses. Fcp. 8vo. 5s.
Virgil's Æneid, translated by Conington. Crown 8vo. 9s.
— Poems, translated into English Prose. Crown 8vo. 9s.

AGRICULTURE, HORSES, DOGS, AND CATTLE.

Dunster's How to Make the Land Pay. Crown 8vo. 5s.
Fitzwygram's Horses and Stables. 8vo. 5s.
Lloyd's The Science of Agriculture. 8vo. 12s.
Loudon's Encyclopædia of Agriculture. 21s.
Miles's Horse's Foot, and How to Keep it Sound. Imperial 8vo. 12s. 6d.
— Plain Treatise on Horse-Shoeing. Post 8vo. 2s. 6d.
— Remarks on Horses' Teeth. Post 8vo. 1s. 6d.
— Stables and Stable-Fittings. Imperial 8vo. 15s.
Nevile's Farms and Farming. Crown 8vo. 6s.
— Horses and Riding. Crown 8vo. 6s.
Steel's Diseases of the Ox, a Manual of Bovine Pathology. 8vo. 15s.
Stonehenge's Dog in Health and Disease. Square crown 8vo. 7s. 6d.
— Greyhound. Square crown 8vo. 15s.
Taylor's Agricultural Note Book. Fcp. 8vo. 2s. 6d.
Ville on Artificial Manures, by Crookes. 8vo. 21s.
Youatt's Work on the Dog. 8vo. 6s.
— — — Horse. 8vo. 7s. 6d.

London: LONGMANS, GREEN, & CO.

SPORTS AND PASTIMES.

The Badminton Library of Sports and Pastimes. Edited by the Duke of Beaufort and A. E. T. Watson. With numerous Illustrations. Crown 8vo. 10s. 6d. each.

> Hunting, by the Duke of Beaufort, &c.
> Fishing, by H. Cholmondeley-Pennell, &c. 2 vols.
> Racing, by the Earl of Suffolk, &c.
> Shooting, by Lord Walsingham, &c. 2 vols.
> *** *Other Volumes in preparation.*

Campbell-Walker's Correct Card, or How to Play at Whist. Fcp. 8vo. 2s. 6d.
Dead Shot (The) by Marksman. Crown 8vo. 10s. 6d.
Francis's Treatise on Fishing in all its Branches. Post 8vo. 15s.
Longman's Chess Openings. Fcp. 8vo. 2s. 6d.
Pole's Theory of the Modern Scientific Game of Whist. Fcp. 8vo. 2s. 6d.
Proctor's How to Play Whist. Crown 8vo. 5s.
Ronalds's Fly-Fisher's Entomology. 8vo. 14s.
Verney's Chess Eccentricities. Crown 8vo. 10s. 6d.
Wilcocks's Sea-Fisherman. Post 8vo. 6s.
Year's Sport (The) for 1885. 8vo. 21s.

ENCYCLOPÆDIAS, DICTIONARIES, AND BOOKS OF REFERENCE.

Acton's Modern Cookery for Private Families. Fcp. 8vo. 4s. 6d.
Ayre's Treasury of Bible Knowledge. Fcp. 8vo. 6s.
Brande's Dictionary of Science, Literature, and Art. 3 vols. medium 8vo. 63s.
Cabinet Lawyer (The), a Popular Digest of the Laws of England. Fcp. 8vo. 9s.
Cates's Dictionary of General Biography. Medium 8vo. 28s.
Doyle's The Official Baronage of England. Vols. I.-III. 3 vols. 4to. £5. 5s.; Large Paper Edition, £15. 15s.
Gwilt's Encyclopædia of Architecture. 8vo. 52s. 6d.
Keith Johnston's Dictionary of Geography, or General Gazetteer. 8vo. 42s.
M'Culloch's Dictionary of Commerce and Commercial Navigation. 8vo. 63s.
Maunder's Biographical Treasury. Fcp. 8vo. 6s.
— Historical Treasury. Fcp. 8vo. 6s.
— Scientific and Literary Treasury. Fcp. 8vo. 6s.
— Treasury of Bible Knowledge, edited by Ayre. Fcp. 8vo. 6s.
— Treasury of Botany, edited by Lindley & Moore. Two Parts, 12s.
— Treasury of Geography. Fcp. 8vo. 6s.
— Treasury of Knowledge and Library of Reference. Fcp. 8vo. 6s.
— Treasury of Natural History. Fcp. 8vo. 6s.
Quain's Dictionary of Medicine. Medium 8vo. 31s. 6d., or in 2 vols. 34s.
Reeve's Cookery and Housekeeping. Crown 8vo. 7s. 6d.
Rich's Dictionary of Roman and Greek Antiquities. Crown 8vo. 7s. 6d.
Roget's Thesaurus of English Words and Phrases. Crown 8vo. 10s. 6d.
Ure's Dictionary of Arts, Manufactures, and Mines. 4 vols. medium 8vo. £7. 7s.
Willich's Popular Tables, by Marriott. Crown 8vo. 10s.

London: LONGMANS, GREEN, & CO.

A SELECTION
OF
EDUCATIONAL WORKS.

TEXT-BOOKS OF SCIENCE.

Abney's Treatise on Photography. Fcp. 8vo. 3s. 6d.
Anderson's Strength of Materials. 3s. 6d.
Armstrong's Organic Chemistry. 3s. 6d.
Ball's Elements of Astronomy. 6s.
Barry's Railway Appliances. 3s. 6d.
Bauerman's Systematic Mineralogy. 6s.
— Descriptive Mineralogy. 6s.
Bloxam and Huntington's Metals. 5s.
Glazebrook's Physical Optics. 6s.
Glazebrook and Shaw's Practical Physics. 6s.
Gore's Art of Electro-Metallurgy. 6s.
Griffin's Algebra and Trigonometry. 3s. 6d. Notes and Solutions, 3s. 6d.
Jenkin's Electricity and Magnetism. 3s. 6d.
Maxwell's Theory of Heat. 3s. 6d.
Merrifield's Technical Arithmetic and Mensuration. 3s. 6d. Key, 3s. 6d.
Miller's Inorganic Chemistry. 3s. 6d.
Preece and Sivewright's Telegraphy. 5s.
Rutley's Study of Rocks, a Text-Book of Petrology. 4s. 6d.
Shelley's Workshop Appliances. 4s. 6d.
Thomé's Structural and Physiological Botany. 6s.
Thorpe's Quantitative Chemical Analysis. 4s. 6d.
Thorpe and Muir's Qualitative Analysis. 3s. 6d.
Tilden's Chemical Philosophy. 3s. 6d. With Answers to Problems. 4s. 6d.
Unwin's Elements of Machine Design. 6s.
Watson's Plane and Solid Geometry. 3s. 6d.

THE GREEK LANGUAGE.

Bloomfield's College and School Greek Testament. Fcp. 8vo. 5s.
Bolland & Lang's Politics of Aristotle. Post 8vo. 7s. 6d.
Collis's Chief Tenses of the Greek Irregular Verbs. 8vo. 1s.
— Pontes Graeci, Stepping-Stone to Greek Grammar. 12mo. 3s. 6d.
— Praxis Graeca, Etymology. 12mo. 2s. 6d.
— Greek Verse-Book, Praxis Iambica. 12mo. 4s. 6d.
Farrar's Brief Greek Syntax and Accidence. 12mo. 4s. 6d.
— Greek Grammar Rules for Harrow School. 12mo. 1s. 6d.
Hewitt's Greek Examination-Papers. 12mo. 1s. 6d.
Isbister's Xenophon's Anabasis, Books I. to III. with Notes. 12mo. 3s. 6d.
Jerram's Graecè Reddenda. Crown 8vo. 1s. 6d.

London: LONGMANS, GREEN, & CO.

Kennedy's Greek Grammar. 12mo. 4s. 6d.
Liddell & Scott's English-Greek Lexicon. 4to. 36s.; Square 12mo. 7s. 6d.
Linwood's Sophocles, Greek Text, Latin Notes. 4th Edition. 8vo. 16s.
Mahaffy's Classical Greek Literature. Crown 8vo. Poets, 7s. 6d. Prose Writers, 7s. 6d.
Morris's Greek Lessons. Square 18mo. Part I. 2s. 6d.; Part II. 1s.
Parry's Elementary Greek Grammar. 12mo. 3s. 6d.
Plato's Republic, Book I. Greek Text, English Notes by Hardy. Crown 8vo. 3s.
Sheppard and Evans's Notes on Thucydides. Crown 8vo. 7s. 6d.
Thucydides, Book IV. with Notes by Barton and Chavasse. Crown 8vo. 5s.
Valpy's Greek Delectus, improved by White. 12mo. 2s. 6d. Key, 2s. 6d.
White's Xenophon's Expedition of Cyrus, with English Notes. 12mo. 7s. 6d.
Wilkins's Manual of Greek Prose Composition. Crown 8vo. 5s. Key, 5s.
— Exercises in Greek Prose Composition. Crown 8vo. 4s. 6d. Key, 2s. 6d.
— New Greek Delectus. Crown 8vo. 3s. 6d. Key, 2s. 6d.
— Progressive Greek Delectus. 12mo. 4s. Key, 2s. 6d.
— Progressive Greek Anthology. 12mo. 5s.
— Scriptores Attici, Excerpts with English Notes. Crown 8vo. 7s. 6d.
— Speeches from Thucydides translated. Post 8vo. 6s.
Yonge's English-Greek Lexicon. 4to. 21s.; Square 12mo. 8s. 6d.

THE LATIN LANGUAGE.

Bradley's Latin Prose Exercises. 12mo. 3s. 6d. Key, 5s.
— Continuous Lessons in Latin Prose. 12mo. 5s. Key, 5s. 6d.
— Cornelius Nepos, improved by White. 12mo. 3s. 6d.
— Eutropius, improved by White. 12mo. 2s. 6d.
— Ovid's Metamorphoses, improved by White. 12mo. 4s. 6d.
— Select Fables of Phædrus, improved by White. 12mo. 2s. 6d.
Collis's Chief Tenses of Latin Irregular Verbs. 8vo. 1s.
— Pontes Latini, Stepping-Stone to Latin Grammar. 12mo. 3s. 6d.
Hewitt's Latin Examination-Papers. 12mo. 1s. 6d.
Isbister's Cæsar, Books I.-VII. 12mo. 4s.; or with Reading Lessons, 4s. 6d.
— Cæsar's Commentaries, Books I.-V. 12mo. 3s. 6d.
— First Book of Cæsar's Gallic War. 12mo. 1s. 6d.
Jerram's Latinè Reddenda. Crown 8vo. 1s. 6d.
Kennedy's Child's Latin Primer, or First Latin Lessons. 12mo. 2s.
— Child's Latin Accidence. 12mo. 1s.
— Elementary Latin Grammar. 12mo. 3s. 6d.
— Elementary Latin Reading Book, or Tirocinium Latinum. 12mo. 2s.
— Latin Prose, Palæstra Stili Latini. 12mo. 6s.
— Subsidia Primaria, Exercise Books to the Public School Latin Primer. I. Accidence and Simple Construction, 2s. 6d. II. Syntax, 3s. 6d.
— Key to the Exercises in Subsidia Primaria, Parts I. and II. price 5s.
— Subsidia Primaria, III. the Latin Compound Sentence. 12mo. 1s.
— Curriculum Stili Latini. 12mo. 4s. 6d. Key, 7s. 6d.
— Palæstra Latina, or Second Latin Reading Book. 12mo. 5s.

London: LONGMANS, GREEN, & CO.

A Selection of Educational Works. 15

Millington's Latin Prose Composition. Crown 8vo. 3s. 6d.
— Selections from Latin Prose. Crown 8vo. 2s. 6d.
Moody's Eton Latin Grammar. 12mo. 2s. 6d. The Accidence separately, 1s.
Morris's Elementa Latina. Fcp. 8vo. 1s. 6d. Key, 2s. 6d.
Parry's Origines Romanæ, from Livy, with English Notes. Crown 8vo. 4s.
The Public School Latin Primer. 12mo. 2s. 6d.
— — — — Grammar, by Rev. Dr. Kennedy. Post 8vo. 7s. 6d.
Prendergast's Mastery Series, Manual of Latin. 12mo. 2s. 6d.
Rapier's Introduction to Composition of Latin Verse. 12mo. 3s. 6d. Key, 2s. 6d.
Sheppard and Turner's Aids to Classical Study. 12mo. 5s. Key, 6s.
Valpy's Latin Delectus, improved by White. 12mo. 2s. 6d. Key, 3s. 6d.
Virgil's Æneid, translated into English Verse by Conington. Crown 8vo. 9s.
— Works, edited by Kennedy. Crown 8vo. 10s. 6d.
— — translated into English Prose by Conington. Crown 8vo. 9s.
Walford's Progressive Exercises in Latin Elegiac Verse. 12mo. 2s. 6d. Key, 5s.
White and Riddle's Large Latin-English Dictionary. 1 vol. 4to. 21s.
White's Concise Latin-Eng. Dictionary for University Students. Royal 8vo. 12s.
— Junior Students' Eng.-Lat. & Lat.-Eng. Dictionary. Square 12mo. 5s.
Separately { The Latin-English Dictionary, price 3s.
{ The English-Latin Dictionary, price 3s.
Yonge's Latin Gradus. Post 8vo. 9s.; or with Appendix, 12s.

WHITE'S GRAMMAR-SCHOOL GREEK TEXTS.

Æsop (Fables) & Palæphatus (Myths). 32mo. 1s.
Euripides, Hecuba. 2s.
Homer, Iliad, Book I. 1s.
— Odyssey, Book I. 1s.
Lucian, Select Dialogues. 1s.
Xenophon, Anabasis, Books I. III. IV. V. & VI. 1s. 6d. each; Book II. 1s.; Book VII. 2s.

Xenophon, Book I. without Vocabulary. 3d.
St. Matthew's and St. Luke's Gospels. 2s. 6d. each.
St. Mark's and St. John's Gospels. 1s. 6d. each.
The Acts of the Apostles. 2s. 6d.
St. Paul's Epistle to the Romans. 1s. 6d.

The Four Gospels in Greek, with Greek-English Lexicon. Edited by John T. White, D.D. Oxon. Square 32mo. price 5s.

WHITE'S GRAMMAR-SCHOOL LATIN TEXTS.

Cæsar. Gallic War, Books I. & II. V. & VI. 1s. each. Book I. without Vocabulary, 3d.
Cæsar, Gallic War, Books III. & IV. 9d. each.
Cæsar, Gallic War, Book VII. 1s. 6d.
Cicero, Cato Major (Old Age). 1s. 6d.
Cicero, Lælius (Friendship). 1s. 6d.
Eutropius, Roman History, Books I. & II. 1s. Books III. & IV. 1s.
Horace, Odes, Books I. II. & IV. 1s. each.
Horace, Odes, Book III. 1s. 6d.
Horace, Epodes and Carmen Seculare. 1s.

Nepos, Miltiades, Simon, Pausanias, Aristides. 9d.
Ovid. Selections from Epistles and Fasti. 1s.
Ovid, Select Myths from Metamorphoses. 9d.
Phædrus, Select Easy Fables.
Phædrus, Fables, Books I. & II. 1s.
Sallust, Bellum Catilinarium. 1s. 6d.
Virgil, Georgics, Book IV. 1s.
Virgil, Æneid, Books I. to VI. 1s. each. Book I. without Vocabulary, 3d.
Virgil, Æneid, Books VII. VIII. X. XI. XII. 1s. 6d. each.

London: LONGMANS, GREEN, & CO.

THE FRENCH LANGUAGE.

Albités's How to Speak French. Fcp. 8vo. 5s. 6d.
— Instantaneous French Exercises. Fcp. 2s. Key, 2s.
Cassal's French Genders. Crown 8vo. 3s. 6d.
Cassal & Karcher's Graduated French Translation Book. Part I. 3s. 6d.
 Part II. 5s. Key to Part I. by Professor Cassal, price 5s.
Contanseau's Practical French and English Dictionary. Post 8vo. 3s. 6d.
— Pocket French and English Dictionary. Square 18mo. 1s. 6d.
— Premières Lectures. 12mo. 2s. 6d.
— First Step in French. 12mo. 2s. 6d. Key, 3s.
— French Accidence. 12mo. 2s. 6d.
— — Grammar. 12mo. 4s. Key, 3s.
Contanseau's Middle-Class French Course. Fcp. 8vo. :—

Accidence, 8d.	French Translation-Book, 8d.
Syntax, 8d.	Easy French Delectus, 8d.
French Conversation-Book, 8d.	First French Reader, 8d.
First French Exercise-Book, 8d.	Second French Reader, 8d.
Second French Exercise-Book, 8d.	French and English Dialogues, 8d.

Contanseau's Guide to French Translation. 12mo. 3s. 6d. Key, 3s. 6d.
— Prosateurs et Poëtes Français. 12mo. 5s.
— Précis de la Littérature Française. 12mo. 3s. 6d.
— Abrégé de l'Histoire de France. 12mo. 2s. 6d.
Féval's Chouans et Bleus, with Notes by C. Sankey, M.A. Fcp. 8vo. 2s. 6d.
Jerram's Sentences for Translation into French. Cr. 8vo. 1s. Key, 2s. 6d.
Prendergast's Mastery Series, French. 12mo. 2s. 6d.
Souvestre's Philosophe sous les Toits, by Stiévenard. Square 18mo. 1s. 6d.
Stepping-Stone to French Pronunciation. 18mo. 1s.
Stiévenard's Lectures Françaises from Modern Authors. 12mo. 4s. 6d.
— Rules and Exercises on the French Language. 12mo. 3s. 6d.
Tarver's Eton French Grammar. 12mo. 6s. 6d.

THE GERMAN LANGUAGE.

Blackley's Practical German and English Dictionary. Post 8vo. 3s. 6d.
Buchheim's German Poetry, for Repetition. 18mo. 1s. 6d.
Collis's Card of German Irregular Verbs. 8vo. 2s.
Fischer-Fischart's Elementary German Grammar. Fcp. 8vo. 2s. 6d.
Just's German Grammar. 12mo. 1s. 6d.
— German Reading Book. 12mo. 3s. 6d.
Longman's Pocket German and English Dictionary. Square 18mo. 2s. 6d.
Naftel's Elementary German Course for Public Schools. Fcp. 8vo.

German Accidence. 9d.	German Prose Composition Book. 9d.
German Syntax. 9d.	First German Reader. 9d.
First German Exercise-Book. 9d.	Second German Reader. 9d.
Second German Exercise-Book. 9d.	

Prendergast's Mastery Series, German. 12mo. 2s. 6d.
Quick's Essentials of German. Crown 8vo. 3s. 6d.
Selss's School Edition of Goethe's Faust. Crown 8vo. 5s.
— Outline of German Literature. Crown 8vo. 4s. 6d.
Wirth's German Chit-Chat. Crown 8vo. 2s. 6d.

London: LONGMANS, GREEN, & CO.

www.ingramcontent.com/pod-product-compliance
Lightning Source LLC
Chambersburg PA
CBHW030359230426
43664CB00007BB/668